NEW KEY GEOGRAPHY

Foundations

Nelson Thornes

DAVID WAUGH
AND TONY BUSHELL

Key Geography Foundations published in 1991 and *New Edition* in 1996 by Stanley Thornes (Publishers) Ltd.
Key Geography New Foundations (Third edition) published in 2001 by Nelson Thornes Ltd.

This edition published in 2006 by:
Nelson Thornes Ltd
Delta Place
27 Bath Road
CHELTENHAM
GL53 7TH
United Kingdom

08 09 10 / 10 9 8 7 6

A catalogue record for this book is available from the British Library

ISBN 978 0 7487 9701 1

Illustrations by Kathy Baxendale, Nick Hawken, Gordon Lawson, Angela Lumley, Richard Morris, David Russell Illustration, Tim Smith, John Yorke
Edited by Katherine James
Photo research by Penni Bickle and Julia Hanson; research for new edition by Sue Sharp
Original design by Hilary Norman
Page make-up by Viners Wood Associates

Printed in China by Midas Printing International Ltd.

Acknowledgements

Cover photos: Waterfall, Loch Awe, Scotland by Joe Cornish/ Digital Vision LL (NT) (left); Jakarta, Indonesian floods by Crack/ Reuters/ Corbis (centre); Foster building ('Gherkin'), London by John Begg Photography/ Photographers Direct (right). Title page photo: James Lauritz/ Digital Vision © (NT).

Actionplus/ Glyn Kirk: 22A; Airphotos: 63C; Alamy/ Aerofilms: 64C; Alamy/ Doug Houghton: 139 (bottom right); Alamy/ JLImages: 20A (left); Alamy/ David Lyons: 103B; Alamy/ Mira: 20A (right); Alamy/ Photofusion/ Dorothy Burrows: 130C (bottom left); Alamy/ Pictures Colour Library: 72B; Alamy/ Hugh Sitton Photography: 80B (left); Ardea/ Bob Gibbons: 130C (bottom middle); Associated Press: 5D; British Tourist Authority/ britainonview.com: 118A; Tony Bushell: 24D (middle left), 27C, 85D, 119D, 139 (middle, top right); Collections/ Liz Staves: 33E; Connors/ Nigel Bowles: 48; Construction Photography: 57 (middle), 130C (top middle); Corbis/ Alberto Garcia: 5B; Corbis/ John Heseltine: 139 (left); Corbis/ Helen King: 57 (left); Corbis/ Reuters/ Yves Herman: 96B; Corbis/ Reuters/ Rafiqur Rahman: 52 (bottom); Corbis/ Patrick Robert: 99C; Corel 777 (NT): 23C; Digital Vision 15 (NT): 40A; Disasters Emergency Committee: 97E; Edifice: 71 (2); Empics/ AP/ Karim Khamzin: 89B; Empics/ AP/ Pavel Rahman: 40B; Empics/ AP/ Vincent Thian: 96C; Empics/ AP/ Apichart Weerawong: 94B (bottom right); Empics/ AP/ Ed Wray: 95B (bottom); Empics/ Razali Nordin: 94B (top left); Empics/ PA/ John Giles: 41D; Empics/ Vincent Thian: 95B (top); Environment Agency: 51B; Environmental Images: 50A (3); Epicscotland.com/ Ashley Coombes: 21A (right); Eye Ubiquitous: 50A (4 & 5), 74B, 83C; Getmapping, supplied by Bluesky International Ltd: 120A (left); Getty Images: 21A (left); Getty Images/ AFP/ Farjana K Godhuly: 55B; Getty Images/ AFP/ John Russell: 92B; Getty Images/ AFP/ Dibyangshu Sarkar: 94B (bottom left); Getty Images/ AFP/ Prakash Singh: 101 (bottom left); Getty Images/ Paula Bronstein: 100; Getty Images/ Paul Scholey: 102A; Hutchison Library: 44C, 71 (3), 111E; James Davis Travel Photography: 106C; Mary Jeffries: 24D (left); John Birdsall Photography: 80B (right), 112B (middle & right); Landscape Only: 12B; Lonely Planet Images/ Neil Setchfield: 60A; Magnum Photos/ Ian Berry: 52 (top); Caroline Malatesta/ Birmingham News, Alabama: 88A; Metrolink: 83C; NERC Satellite Receiving Station, University of Dundee: 32B, 34C; NHPA: 130C (top left); Panos Pictures: 54A (top left & bottom right); Panos Pictures/ Fredrik Naumann: 96A; Penni Bickle: 50A (2), 74A; Photofusion/ Jacky Chapman: 120 (top); Photofusion/ Stan Gamester: 118B; Photofusion/ Bob Watkins: 61C; Photolibrary.com: 130C (top right); Photolibrary.com/ Mark Hamblin: 57 (right); Photolibrary.com/ ifa-Bilderteam Images: 4A; Photolibrary.com/ Jon Arnold Images: 4C; Jim Reed/ Digital Vision WW (NT): 22B; Rex Features/ Michael Dunlea: 94 (top right); Rex Features/ Brian Harris: 23D; Rex Features/ Masatoshi Okauchi: 101 (top right); Rex Features/ Jim Pickerell: 112B (left); Rex Features/ Colin Shepherd: 41C; Rex Features/ Sipa Press: 101 (top left); Science Photo Library: 114A; Skyscan.co.uk: 64A, 130C (bottom right); Skyscan.co.uk/ K Allen: 13D; Skyscan.co.uk/ APS: 103D; Skyscan.co.uk/ CLI: 64B; Skyscan.co.uk/ LAPL: 61B; Still Pictures: 50A (1), 54A (top middle & top left), 55D, 77C, 80A; Tower Hamlets Local History Library: 72A; John Walmsley: 18A; Dr A C Waltham: 46A; Simon Warner: 76A, 106B, 141C; David Waugh: 71B (1); R D Whyman: 24D (right); Ken Woodley: 24D (middle right); York and County Press: 50A (6).

Maps produced from Ordnance Survey mapping with the permission of Ordnance Survey on behalf of HMSO. © Crown copyright (2006). All rights reserved. Ordnance Survey Licence number 100017284: 65D (Landranger 202); 116A and B (Explorer 316); 124B (Explorer 277); 125D, 131D, 140B and inside back cover (Landranger 154).

Screenshot and data (121B and C): Source: National Statistics website: www.statistics.gov.uk. Crown copyright material is reproduced with the permission of the Controller of HMSO.

The cartographic information which appears in this publication (124A and C) was supplied by © Automobile Association Developments Ltd 2005 LIC038/05. All rights reserved. Includes mapping data supplied by © Ordnance Survey, Crown Copyright 2005. All rights reserved. Licence 399221.

Screenshots (125D all) reproduced with kind permission of Anquet Technology Ltd www.anquet.co.uk, and Getmapping, supplied by Bluesky International Ltd (left).

Every effort has been made to contact copyright holders and we apologise if any have been overlooked.

Contents

1 What is geography?   4

Your passport to the world   4
What is physical geography?   6
What is human geography?   8
What is environmental geography?   10
How do we study geography?   12
How can we find out where places are?   14
How can we use graphs in geography?   16
How can we use computers in geography?   18
What is the value and use of geography?   20

2 Weather and climate   22

How can the weather affect us?   22
How might you observe and record the weather?   24
How can local features affect temperature and wind?   26
What is Britain's weather?   28
How does it rain?   30
Forecasting the weather – anticyclones   32
Forecasting the weather – depressions   34
The weather enquiry   36

3 River flooding   40

Why is flooding a problem?   40
How does the water cycle work?   42
What is a river basin?   44
Where are the world's most important rivers?   45
What causes a river to flood?   46
Floods in the UK, 2000   48
How does the UK cope with floods?   50
Floods in Bangladesh, 2004   52
How does Bangladesh cope with floods?   54
How can the risk of flooding be reduced?   56
The river flooding enquiry   58

4 Settlement   60

What are settlements like?   60
How were the sites for early settlements chosen?   62
What different settlement patterns are there?   64
How do settlements change with time?   66
What are the benefits and problems of settlement growth?   68
Why are there different land use patterns in towns?   70
Why does land use in towns change?   72

Where do we shop?   74
How has shopping changed?   76
How does internet shopping affect us?   78
Traffic in urban areas – why is it a problem?   80
Traffic in urban areas – is there a solution?   82
Where should the by-pass go?   84
The settlement enquiry   86

5 The Indian Ocean tsunami   88

The world's worst natural disaster?   88
What caused the tsunami?   90
How did the tsunami affect different countries?   92
What were the effects of the tsunami?   94
How did the world help?   96
How can the tsunami danger be reduced?   98
The tsunami enquiry   100

6 The United Kingdom   102

What is the UK like?   102
Where is the UK?   104
What is the UK?   105
What are the UK's main physical features?   106
How is the UK divided up?   108
Where do people in the UK come from?   110
How well off is the UK?   112
What are the regional differences in the UK?   114
How can maps help us describe the local area?   116
What is it like where you live?   118
How can the internet help?   120
The UK enquiry   122

7 Map skills   124

How can we use maps?   124
How can we show direction?   126
How can we measure distance?   128
How do we use map symbols?   130
What are grid references?   132
How do we use six figure grid references?   134
How is height shown on a map?   136
How do contours show height and relief?   138
How can we describe routes?   140

Glossary and Index   142

1 What is geography?

Your passport to the world

What is this unit about?

This unit explains what is meant by geography and looks at some of the ways we study the subject. It also shows what you will learn from geography and how it can be valuable to you in later life.

In this unit you will learn about:

- differences between physical, human and environmental geography
- how to find places on a map
- how to use maps and photos to describe places
- how to understand and appreciate geography
- the value and use of geography.

Why is geography an important subject?

Geography is about people and places. It helps us understand our world and makes it a more interesting place in which to live. It helps us make sense of news events and what is going on around us. It also helps us understand ways of life that are different from our own and makes travel and meeting people more exciting.

Learning geography can also be of benefit to you in the future.

- It can give you an interest in people and places.
- It can introduce you to a variety of hobbies.
- It can give you job opportunities in a variety of interesting careers.

Put all this together and geography can be your passport to the world.

A Monument Valley

B Mount Pinatubo

C New York

D Bangladesh

◆ Of the places shown in these photos, which
 – would you most like to visit
 – is the most different from where you live
 – is the most attractive
 – is the busiest
 – looks the most dangerous
 – would you like to know more about?

What are your reasons for your answers?

What is physical geography?

Geography is the study of the earth's natural features. It is also about people and places and how they affect each other. Geography can help us to understand our world and, hopefully, to make it a better place in which to live.

One of the best ways to learn about geography is to ask questions. You will notice that most pages in this book start with a key question. For example:

◆ What or where is it?
◆ What is it like?
◆ How did it get like this?
◆ How is it changing?
◆ What might be the effects of these changes?
◆ What do I think about them?

There are three main parts to geography. These are physical, human and environmental.

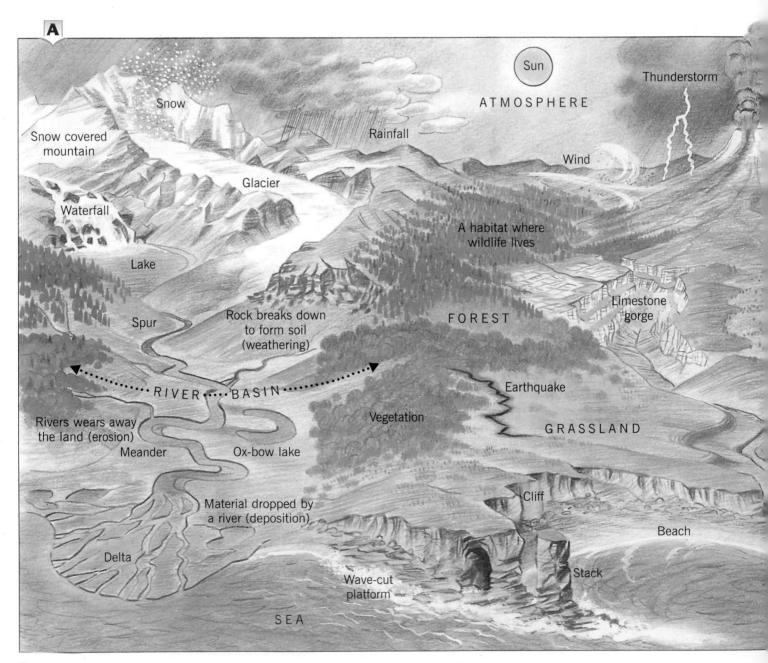

Physical geography is the study of the earth's natural features. It is about the land and the sea and the atmosphere around us.

The **atmosphere** is the air around the earth. Changes in temperature, rainfall and pressure give us our **weather** and **climate**. Climate changes between seasons and from year to year. Different parts of the world have different climates.

Landforms are natural features formed by rivers, the sea, ice and volcanoes. They are continually changing as they are worn away in some places and built up in others.

Most changes in physical geography happen very slowly. Sometimes when sudden changes happen, they cause **hazards** such as storms, floods, drought, volcanic eruptions and earthquakes.

The earth's surface is made up of many different kinds of rock. Where these rocks break up into small pieces, they form soil. Plants grow in this soil and cover most of the earth's land surface.

An **ecosystem** is a community of plants and animals whose lives are closely linked to each other and to the climate and soil of the area in which they grow.

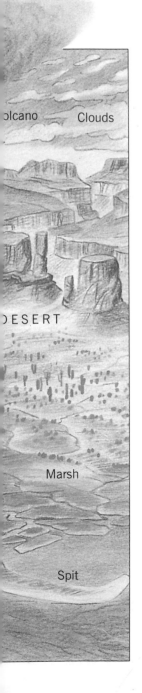

Volcano   Clouds

DESERT

Marsh

Spit

Activities

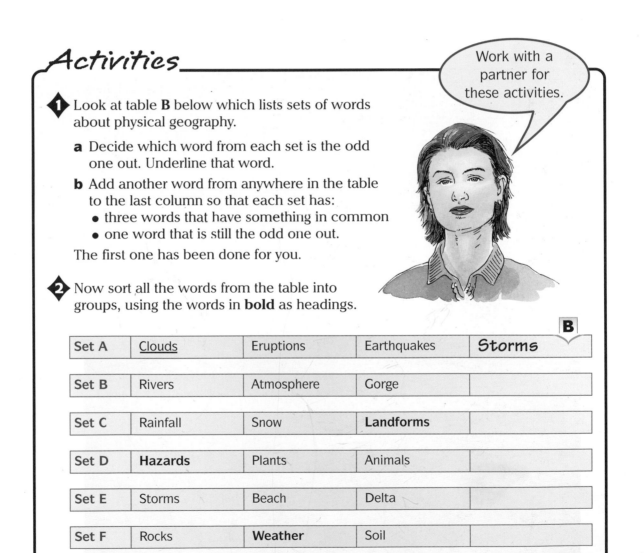

Work with a partner for these activities.

1 Look at table **B** below which lists sets of words about physical geography.

   a Decide which word from each set is the odd one out. Underline that word.

   b Add another word from anywhere in the table to the last column so that each set has:
    • three words that have something in common
    • one word that is still the odd one out.

   The first one has been done for you.

2 Now sort all the words from the table into groups, using the words in **bold** as headings.

B

Set A	<u>Clouds</u>	Eruptions	Earthquakes	*Storms*
Set B	Rivers	Atmosphere	Gorge	
Set C	Rainfall	Snow	**Landforms**	
Set D	**Hazards**	Plants	Animals	
Set E	Storms	Beach	Delta	
Set F	Rocks	**Weather**	Soil	
Set G	Drought	Floods	**Ecosystems**	
Set H	Temperature	Wind	Waterfall	
Set I	Spits	Stack	Climate	

What is human geography?

This is the study of where and how people live.

Population geography looks at the spread (distribution) of people over the earth's surface. It tries to explain why some parts of the world have many people living there while other parts have very few.

It studies areas where the numbers of people living there are growing rapidly, and looks at the problems that come from this growth. It suggests reasons why

people move from one area or country to another (**migration**). It looks at how such movements lead to people having different customs, religious beliefs and ways of living.

Settlement geography is about where people live. It looks at why settlements grow up in a particular place, and why some remain small in size (villages) while others may grow into very large **urban** centres (cities). It describes problems that go with living in

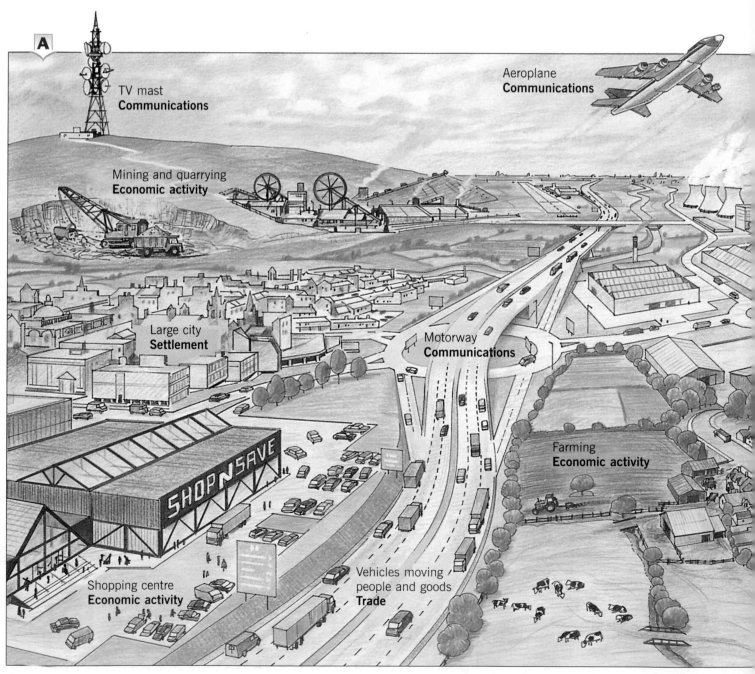

A

TV mast
Communications

Aeroplane
Communications

Mining and quarrying
Economic activity

Large city
Settlement

Motorway
Communications

Farming
Economic activity

Shopping centre
Economic activity

Vehicles moving
people and goods
Trade

very small places as well as those of very large ones. It looks at how land is used in cities and how this use can change over time.

Communications describes the methods of transport by which people may move about – to work, to school, to the shops and for recreation and holidays. It also includes the movement of goods (**trade**) and information, such as conversations on the telephone and programmes on the television.

Economic geography (economic activity) looks at how people try to earn a living. It is about industry, about jobs and about wealth. It is usually divided into three types. These include farming (a **primary** activity),

making things in a factory (a **secondary** activity), or looking after people (a **tertiary** activity or service). It looks at why some activities are only found in certain places and why some parts of the world are richer and more developed than others.

People are very concerned with their **quality of life**. This might be how happy or content they are, the amount of money they have, or how much they like living and working in a particular area. The quality of life may differ greatly both within a country and between countries.

Activities

On a copy of this *Kriss Kross Puzzle*, solve the puzzle by fitting 11 words and phrases into their correct position in the diagram. The 11 words and phrases are those written in **bold** on these two pages. Two have been done to help you get started.

QUALITYOFLIFE

MIGRATION

dustry
economic activity

mall village
Settlement

What is environmental geography?

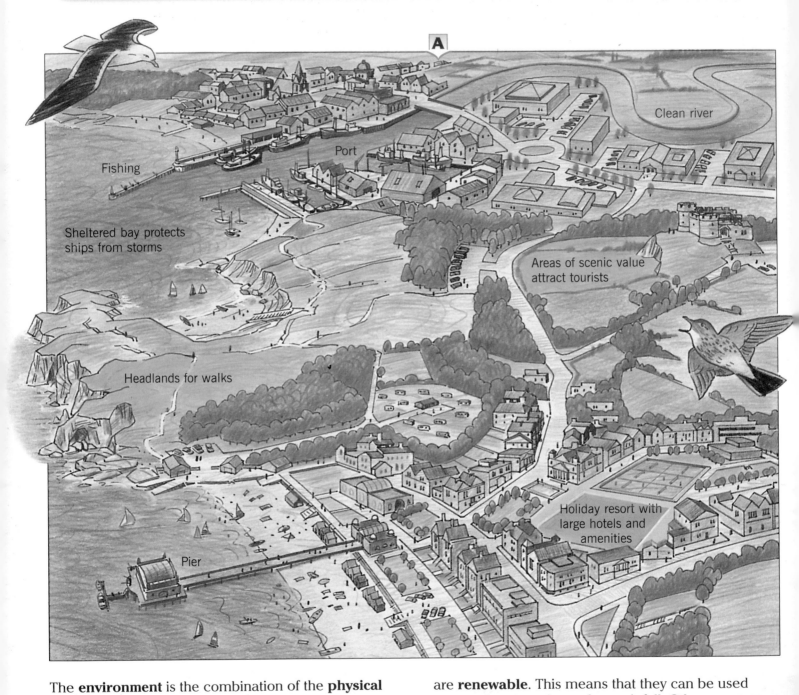

A

Clean river

Port

Fishing

Sheltered bay protects ships from storms

Areas of scenic value attract tourists

Headlands for walks

Holiday resort with large hotels and amenities

Pier

The **environment** is the combination of the **physical** (natural) environment of climate, landforms, soils and vegetation, and the **human** environment which includes settlements and economic activities. It is the study of the surroundings in which people, plants and animals live.

The environment includes natural **resources** such as coal and iron ore, soils, forests and water. These are used to meet human needs. Some of these resources are **renewable**. This means that they can be used over and over again, such as rainfall. Others are **non-renewable** and can only be used once, such as coal. Sometimes people use these resources to their advantage. For example they use water for drinking purposes, iron ore in industry, and landforms such as islands or lakes for leisure. People often misuse these resources by using them up (minerals), by destroying them (soils, forests) or polluting them (rivers, seas and the air).

Different environments have different qualities and different uses. Each needs to be **protected** and carefully **managed**, like National Parks and the reserves of oil. Many environments have been damaged in the past. Those which have, such as mining areas, rivers and the older parts of some cities, need to be improved.

There is now a growing concern over the **quality of the environment** and how it may be **conserved** while at the same time being made as useful as possible.

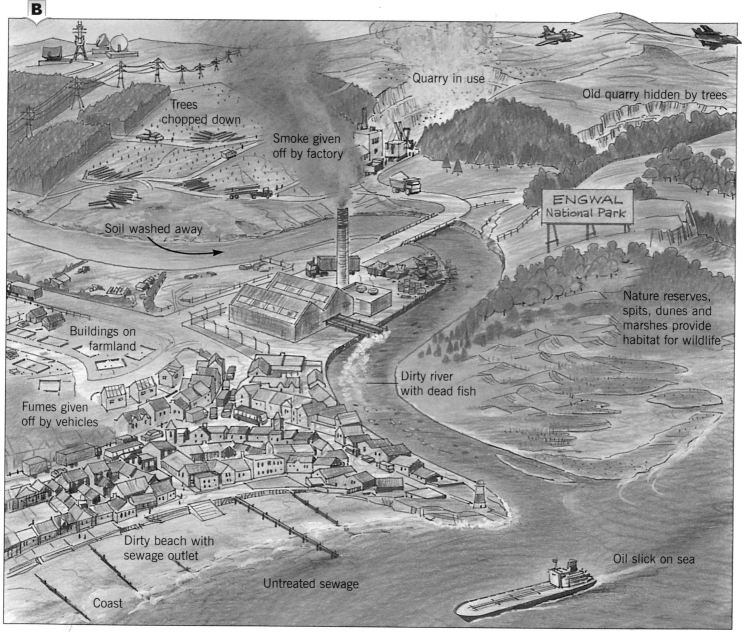

B

Quarry in use

Old quarry hidden by trees

Trees chopped down

Smoke given off by factory

ENGWAL National Park

Soil washed away

Nature reserves, spits, dunes and marshes provide habitat for wildlife

Buildings on farmland

Dirty river with dead fish

Fumes given off by vehicles

Dirty beach with sewage outlet

Oil slick on sea

Untreated sewage

Coast

Activities

1 a Make a copy of the table on the right.

   b List the features shown in drawing **A** in the two columns. Two of these features have already been named for you.

2 In what ways has the area shown in drawing **B** been

   a polluted or destroyed

   b protected?

Physical (natural) environment	Human environment
River	Town

How do we study geography?

Geographers need to know about **places**. They should be able to describe where a place is found (located), why it is there (site) and what it might be like to live or work there. Places can include physical features like rivers, mountains and deserts. Places can also be made by humans, e.g. houses, cities and roads.

Places can vary in **size**. Just as the classroom is a place in a school, so is the school a place in a town, the town is a place in a country, and the country a place in the world.

Diagram **A** is a plan of a classroom. A classroom is a **place** in a school. If the classroom is neat and tidy (like your bedroom at home!) everything will have its own place. The teacher will have a desk, atlases will be kept on a shelf or in a cupboard, and pens and pencils in a box.

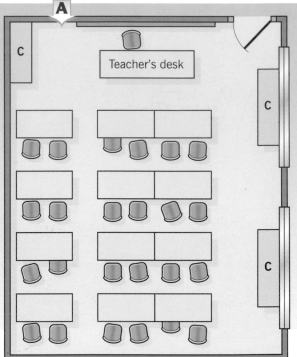

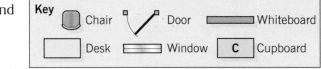

Key

Chair   Door   Whiteboard

Desk   Window   **C** Cupboard

In the distance are some low **hills** which are partly covered in **trees**. There is a **village** in the centre of the photo, with a **church** and several **buildings**. The church has a **tower**. A small **road** passes through the village. In front of the village is a **river** which is crossed by a **bridge**. The land around the village consists of **fields** in which **grass** appears to be growing. The fields are separated by **stone walls** and a few **deciduous trees**. The photo was taken on a **sunny** day in **summer** in the **country**.

How do we describe what a place looks like?

Although no two places in the world are exactly the same, they may have similarities. We have to learn how to describe one place so that we can compare it with a second place. We can show how they are similar to or different from each other.

The best way is to use a photo, possibly from a book or a magazine. When writing a description it is important to pick out **key words**. Key words are important ones to learn and to remember. In the description below photo **B**, the key words have been written in **bold** type so that they are easier to pick out.

We can also describe a place by drawing a labelled (annotated) fieldsketch. Sketch **C** on page 13 is drawn from photo **B**. The labels on the fieldsketch are very similar to the key words in the written description.

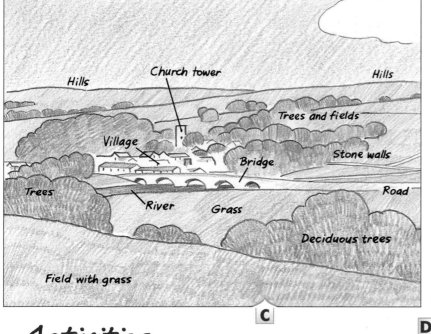

Activities

1 Photo **D** was taken in central London. The key words have been missed out of description **E** and are listed at the end.

Copy out the description in **E**, putting the key words in the correct places.

2 **a** Make a copy of the tables below.

b Complete the table for photo **B** by adding the key words from fieldsketch **C**.

c Complete the table for photo **D** by using the key words listed in description **E**.

Photo B	
Physical features	Human features

Photo D	
Physical features	Human features

3 **a** Which photo has more physical features than human features? Why?

b Which photo has more human features than physical features? Why?

4 Which of the two places shown in photos **B** and **D** would you rather visit or live in? Try to give reasons for your answers.

E

In the foreground is a _____ with a large _____, several small _____ and a _____ _____. In the trees close to the river there is an old _____. The area is packed with _____ and there is very little _____ _____. In the centre of the photo there are several _____ _____ which look like _____. The photo was taken on a _____ day in a big _____.

bridge   castle   offices   city
boats   sunny   river   tall buildings
landing stage   open space   buildings

5 Describe the place where you live, as follows.

a First, pick out a number of key words.

b Second, write a description using the key words.

c Third, draw a simple labelled sketch of one of the main features of the area.

Summary

Geographers study places where people live and those they avoid. Places can be described from photos or by drawing a labelled sketch and underlining key words.

How can we find out where places are?

People often need to know where places are. They need to know this if, for example, they are going shopping or on holiday. Many people, like lorry drivers and ambulance drivers, need to know where places are to do their job. On television we are always hearing about different places, on the news and in other programmes.

Geographers use maps to find out where places are and what they are like. An **atlas** is a book that has maps showing places all around the world, and it is easy to use. The most accurate way to show the whole world is on a **globe**. This is because a globe, being round, shows the actual shape of the earth.

To help us find places, imaginary lines called **latitude** and **longitude** are drawn onto the globe. These are shown in diagram **A**.

A

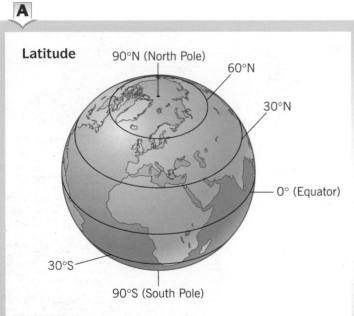

Latitude

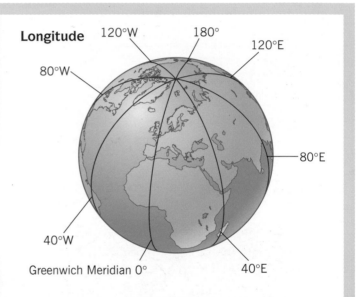

Longitude

Lines of latitude are imaginary lines going around the earth from east to west. They are measured in degrees north or south of the **Equator**. Latitude 0° is called the Equator.

Lines of longitude are imaginary lines going from the North Pole to the South Pole. They are measured from the **Greenwich Meridian**, which passes through London. The Greenwich Meridian is 0°.

It is impossible to draw the earth accurately on a piece of paper. Parts of it will always be either the wrong size or the wrong shape. This is because the earth is round, and a piece of paper is flat. Imagine peeling the skin off an orange and trying to lay it out flat. It cannot be done, because the peel will split and some parts will be pushed out of shape.

One way of drawing the globe as a flat map is shown in map **B**. Some places have been stretched, and others squashed to make them fit.

B

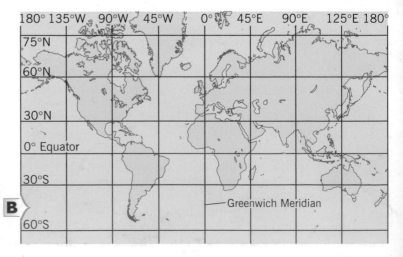

Using an atlas

The **contents** page at the front of the atlas shows on which page each map can be found. The **index** at the back of the book shows exactly where a particular place may be found. The index gives the latitude and longitude of that place to help you find it more easily. Diagram **C** shows you how to use the index of an atlas.

C

Manchester	UK	18	53° 30´N	2° 15´W
Place name	Country	Page	Latitude	Longitude

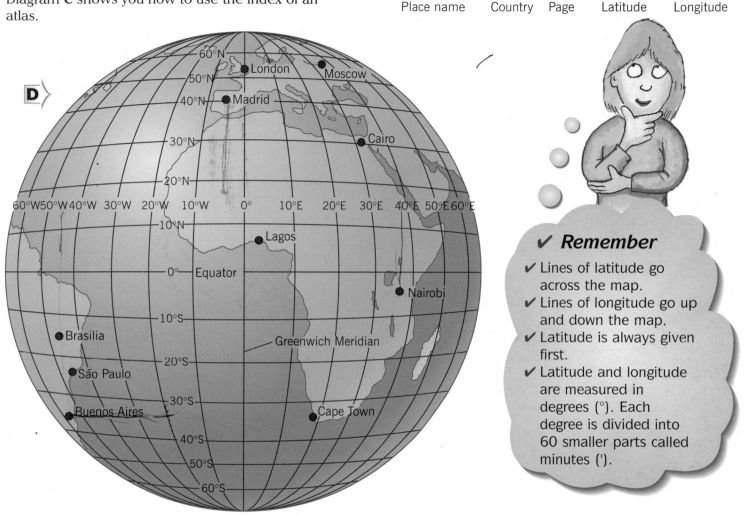

D

✔ **Remember**
✔ Lines of latitude go across the map.
✔ Lines of longitude go up and down the map.
✔ Latitude is always given first.
✔ Latitude and longitude are measured in degrees (°). Each degree is divided into 60 smaller parts called minutes (').

Activities

1 Look at map **D** above. Name the city at each of the following:

a 30°N 31'E **b** 34°S 18'E
c 40°N 4'W **d** 24°S 47'W.

2 Use map **D** to give the latitude and longitude for each of these cities:

a London **b** Lagos
c Moscow **d** Buenos Aires
e Nairobi **f** Brasilia.

3 Find each of the cities below in the index of your atlas. For each one give the country, page number and latitude and longitude.

New York	Tokyo	Sydney	Calcutta

Summary

Maps are useful to people. They help us to find out where places are and what they are like. An atlas shows many places around the world. These places may easily be found using latitude and longitude.

How can we use graphs in geography?

Graphs are diagrams that show information in a clear and simple way. They can be used to describe a situation and show how one thing is related or linked to another.

Graphs are drawn using facts and figures which are called **data**. We can obtain data either by collecting information from fieldwork or by looking it up in a book. Information that we collect ourselves is called **primary data**. Information from other sources is called **secondary data**.

Graphs can either be drawn by hand or on a computer using a spreadsheet program.

✔ Remember

When you draw a graph, it should have:
- ✔ a title to say what it is showing
- ✔ labels along the bottom and up the side to explain what they are showing
- ✔ figures that are plotted very accurately.

Bar graphs **A**

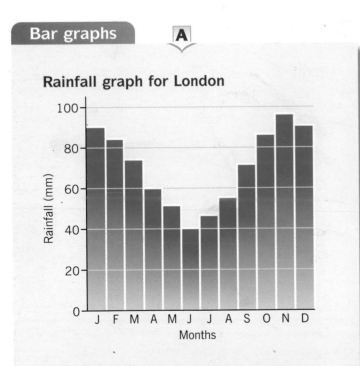

Rainfall graph for London

- ◆ A bar graph is made up of several bars or columns.
- ◆ The bars can be drawn either horizontally across the page or vertically up and down the page.
- ◆ Bar graphs are used to compare different things or quantities.
- ◆ The graph above compares the amount of rain in each month of a year. It shows which parts of the year are wettest.

Line graphs **B**

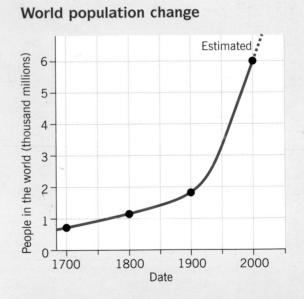

World population change

- ◆ Line graphs show information as a series of points that are joined up to form a line.
- ◆ Line graphs show changes or trends over a period of time. They can also help predict or forecast future changes.
- ◆ The graph above shows that population grew slowly between 1700 and 1800, then very rapidly between 1900 and 2000. The graph also suggests what might happen in the future.

Activities

1 Look at graph **A**.

   a Name the four months with least rainfall.

   b How much rainfall was there in November?

2 Look at graph **B**.

   a What was the population in 1700?

   b What does the graph suggest might happen to population in the future?

3 Look at graph **C**.

   a In which season are fewest holidays taken?

   b What percentage of holidays are taken in summer?

4 Look at graph **D**.

   a How much rainfall gave a river depth of 2 metres?

   b Which point, A, B or C, shows there to be little rainfall and a small amount of water in the river?

5 Which type of graph would be best for:

   a showing the change in the cost of petrol

   b comparing the size of UK cities

   c showing how the land use of an area is divided up?

Summary

Graphs are diagrams used to show data clearly. The four types of graph are the bar graph, line graph, pie graph and scatter graph.

Pie graphs   C

People taking holidays in the UK

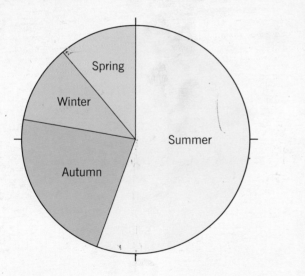

- A pie graph is drawn as a circle which is then divided into several pieces or sectors.
- The whole circle is always equal to 100%.
- Pie graphs show proportions and help us to see how something is divided up.
- The graph above shows that more than half of the people in the UK take their holidays in the summer months.

Scatter graphs   D

Rainfall and river depth

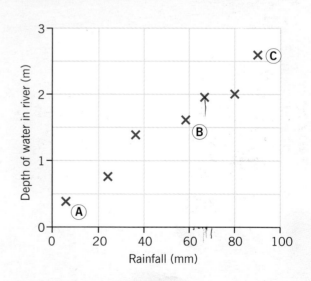

- A scatter graph has data plotted as a number of dots or crosses.
- Scatter graphs are used to see if information about two different things is related or linked.
- The graph above shows the link between rainfall and the amount of water in a river. As rainfall increases, so the amount of water in the river increases.

How can we use computers in geography?

Computers are useful to geographers and can be used in many different ways. At a simple level they can be used to write text, draw diagrams and find out information. At more advanced levels they may be used for such things as forecasting the weather, predicting where and when floods might occur, and researching how earthquakes and volcanic eruptions happen. Computers may also be used to communicate information.

Computers can also be used for teaching geography in the classroom. Images and interactive activities can be shown on whiteboards or via a digital projector for the whole class, or on PCs for individual work. Computers can help to make geography more exciting and interesting.

Some ways that computers, or **Information and Communication Technology (ICT)**, can be used in geography are shown below.

A

A **word processor** can be used to write notes, reports or essays about a topic that you are studying. The text can easily be changed, printed out or saved for later use.

The **internet** is useful for geographical research. It can provide up-to-date statistics and information on almost every geographical topic.

A **spreadsheet** program can be used to draw graphs and diagrams. These show statistical information clearly and are most useful when completing a geographical investigation.

CD-ROMs are like electronic books. They contain large amounts of information about particular topics. Most are interactive and some can be used to simulate, or copy, geographical situations.

Using computers in geography

Desktop publishing (DTP) can be used to present information in a clear and attractive way. It is best used when good design is needed and impact is important.

E-mail is electronic mail. It can be used to communicate with other users quickly and easily. E-mail is excellent for exchanging information and sharing views about a topic or an issue in geography.

The internet is very useful to geographers. It provides almost instant access to people all over the world, and gives up-to-the-minute information on just about every topic and issue in geography.

Much of the information on the internet is updated daily, and it can come in a variety of forms. These include text, statistics, videos, photographs, satellite images, weather charts, diagrams, newspaper reports, eye-witness accounts and even government records.

The internet also allows teachers, pupils and schools from all over the world to communicate with each other, sharing information, ideas and experiences.

Nelson Thornes, the publisher of this book, has developed an internet site especially for Key Geography. It provides resources and information that are directly linked to books in the series. The site also gives website addresses that you might find useful in your studies.

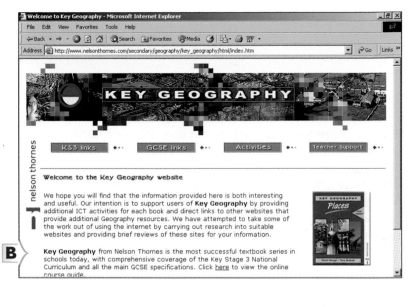

The Key Geography website at:

www.nelsonthornes.com

then click

secondary/geography/key_geography.htm

Activities

1
a Make a larger copy of table **C**. You will need three lines for each use or application.

b Complete the table by adding examples from list **D** below.

c Add an example of your own to each use.

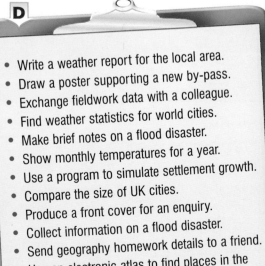

D
- Write a weather report for the local area.
- Draw a poster supporting a new by-pass.
- Exchange fieldwork data with a colleague.
- Find weather statistics for world cities.
- Make brief notes on a flood disaster.
- Show monthly temperatures for a year.
- Use a program to simulate settlement growth.
- Compare the size of UK cities.
- Produce a front cover for an enquiry.
- Collect information on a flood disaster.
- Send geography homework details to a friend.
- Use an electronic atlas to find places in the world.

C Computer use or application	Examples
Word processor	
Graphs or diagrams	
Desktop publishing	
Internet	
E-mail	
CD-ROM	

2
a What does ICT stand for?

b Why do you think it is useful to use ICT in geography?

3
a Log onto the internet and visit the Key Geography website at the address given above.

b Make a list of the pages and features that may be useful to you in your geography studies.

Summary

Computers are important in geography. They give access to information sources, and help in presenting and communicating geographical data.

What is the value and use of geography?

The knowledge and skills that you learn in geography will help you in the future. They will give you an interest in people and places and help you understand what is going on in the world. They will also help you to make more sense of events in the news and enable you to develop your own views and opinions about both local and global issues.

As you will see, one of the best ways to learn about geography is to ask questions. Indeed you will notice that most pages in this book start with a key question and each chapter ends with an enquiry. Learning to ask questions and develop enquiry skills will help you find out things for yourself and make your own decisions.

But that's not all. The knowledge and skills that you learn in geography can also open the door to a variety of interesting and exciting careers. Jobs in travel, town planning, weather forecasting, mapping, journalism and the environment are just some of these.

Physical geography

Physical geography helps you:

◆ learn about the earth's landforms, climate and vegetation

◆ appreciate landscapes and scenery

◆ make sense of your surroundings

◆ be aware of what other places in the world are like

◆ know the causes of natural hazards

◆ know how to prepare for and cope with natural hazards.

Human geography

Human geography helps you:

◆ learn about and appreciate your surroundings

◆ learn what other countries and cities are like

◆ understand population growth and migration

◆ understand ways of life that are different from your own

◆ learn how and why countries are at different stages of development

◆ learn about some of the problems facing our world and how we might solve them.

Activities

1 Make a copy of the table below.

　a List the topics from drawing **B** in the correct columns.

　b Add at least two more topics to each column.

Some examples of topics in geography			
Physical	Human	Environmental	Skills

2 In what ways do you think geography will help *you* in your life? Try to give at least six.

3 Make a list of careers where the knowledge and skills that you learn in geography will be a help. Try to give at least ten.

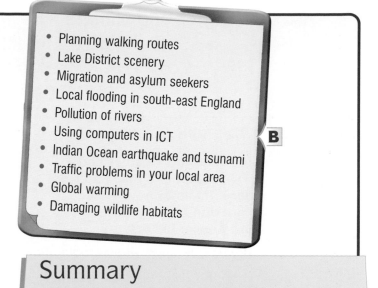

- Planning walking routes
- Lake District scenery
- Migration and asylum seekers
- Local flooding in south-east England
- Pollution of rivers
- Using computers in ICT
- Indian Ocean earthquake and tsunami
- Traffic problems in your local area
- Global warming
- Damaging wildlife habitats

B

Summary

The knowledge and skills that you learn in geography can help you understand our world and will help you in future years.

Environmental geography

Environmental geography helps you:

- learn how we use the earth's natural resources
- understand what happens when we waste resources and damage the environment
- learn that we must live in a sustainable way
- learn how to recycle waste materials and reduce energy consumption
- develop a concern for the environment
- learn about protecting and conserving wildlife and scenery.

Skills in geography

Skills in geography help you:

- read maps and find your way about
- interpret and use graphs and statistics
- use maps and photos to find out what places are like
- learn how to carry out enquiries
- use questions and enquiry skills to find out things for yourself
- use computers (ICT) to find things out and present information.

How can the weather affect us?

What is this unit about?

This unit is about how weather and climate vary from time to time and from place to place. It shows how these variations are due to many different physical and human factors.

In this unit you will learn about:

◆ observing and recording the weather

◆ how local features affect temperature and wind

◆ what causes rain

◆ how weather and climate vary across Britain

◆ anticyclones and depressions

◆ how to forecast the weather.

A Wimbledon

B East Anglia

Why is this weather topic important?

Weather affects our lives in many ways. For example, it affects:

◆ the sort of activities we do

◆ the type of clothes we wear

◆ what we plan to do at the weekend

◆ where and when we go on holiday.

An understanding of the weather helps us make better use of weather forecasts and may even help us make predictions of our own. This can help us make the best use of weather conditions and to avoid the problems that unexpected changes in the weather can bring. This unit will help you to do that.

◆ For each photo, how would
  – the weather affect you
  – you feel about being there
  – you prepare for that weather?

◆ What problems may these types of weather cause?

C Canary Islands

D UK city

How might you observe and record the weather?

Weather can be described as the condition of the air around us over a short period of time. It is about being hot or cold, wet or dry, windy or calm, cloudy or sunny.

Meteorology is a study of the weather. One of the important tasks of meteorologists is to measure and record all the features of the weather every day. Many expensive and complicated instruments are needed to record weather accurately but you can get a good picture of what conditions are like by **observing** (looking around) and using simple equipment.

A

Temperature
This is a measure of how hot or cold it is. You can do this by looking at the clothes that people are wearing. Thermometers are used to measure temperature accurately.

B

Precipitation
Water in the air falls to the ground in one of several forms. Four of these are rain, snow, sleet and hail.

C

Wind speed
This tells us how strong the wind is. We can get a good idea of this by looking at smoke and the trees. The **Beaufort scale** is used to measure wind strength.

D

Cloud type
Cloud comes in many shapes, sizes and heights. Cumulonimbus, cumulus, stratus and cirrus are the most common types.

A — Very cold | Cold | Mild | Warm | Hot

B — 0°C / 0°C | Below 0°C / Above 0°C

C — 0 Calm — Smoke rises vertically | 2 Light breeze — Wind felt on face, leaves rustle | 4 Moderate breeze — Dust and paper lifted, small branches move | 6 Strong breeze — Large branches in motion | 8 Fresh gale — Twigs break off trees

D — Cumulonimbus | Cumulus | Stratus | Cirrus

E

Wind direction
This is the direction **from** which the wind blows. It is measured by a wind vane.

F

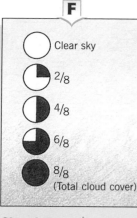

Clear sky
2/8
4/8
6/8
8/8
(Total cloud cover)

Cloud cover
This is the amount of the sky covered by cloud. It is measured in eighths.

G

Visibility
This is the distance that can be seen. It is measured in metres.

H

General weather
This describes the weather in words. Words like rain, snow, showers, fog, mist, thunder, cloudy, fair or sunny are used. Light or heavy can be added to precipitation.

Activities

1 What is weather?

2 **a** Make a copy of diagram **I** on the right.
   b Write the name of the weather feature next to each sketch.

3 Describe how each of the following is measured:

temperature

wind direction

wind strength

cloud cover

4 Make sketches of the four cloud types in **D**. Under each sketch write a cloud description from the following list.
   • Low grey shapeless cloud that forms in layers.
   • High clouds that are wispy, light and featherlike.
   • Dome shaped clouds with dark flat bases.
   • Huge towering clouds that often give showers.

OK

Weather features to be observed and recorded

I

5 Look at the table **J** below which shows what the weather was like on a summer day in Wales.
   a Copy out the table headings.
   b Make your own recording of today's weather. Use the information on these two pages to help you.

EXTRA

1 Keep a record of the weather for a week. Do this at the same time each day. Record your readings in a table.

2 See if you can spot any link between the wind direction and other features of the weather.

J

Day	Temperature	Precipitation	Wind speed	Wind direction	Cloud amount	Cloud type	Weather
Sunday 15 July	Warm	Rain showers	Force 2	Westerly	4/8	Cumulus	Mainly sunny with some rain
Monday 16 July							

Summary

Weather is the day to day condition of the atmosphere. A simple record of the weather may be made by careful observation of what is going on around us.

How can local features affect temperature and wind?

On a fine summer's day, are some of the classrooms in your school hotter than others? When the sun shines or a cold wind blows, is one side of your classroom warmer or colder than the other? On a hot sunny day can you notice a difference in temperature between a dark, tarmac playground and a grassy area like the school field? Are there some sheltered places around your school where you can get out of the wind?

Look at cartoon **A** which shows how different the conditions can be on two sides of a hedge.

Each particular place or site tends to develop its own special climate conditions. When the climate in a small area is different from the general surroundings it is called a **microclimate**. Some of the causes of microclimates are given in **B** below.

A

B

Shelter

Trees, hedges, walls, buildings and even hills can provide shelter from the wind. Wind speed may be reduced and its direction changed. Places sheltered from cold winds will be warmer.

Physical features

Trees provide shade and shelter and are usually cooler than surrounding areas. Water areas such as lakes and seas have a cooling effect and may also produce light winds. Hilltops are usually cool and windy.

Surface

The colour of the ground surface affects warming. Dark surfaces such as tarmac and soil will become warmer than light surfaces such as grass.

Buildings

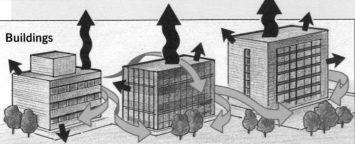

Buildings give off heat that has been stored from the sun during the day or which leaks from their heating systems. Temperatures near buildings may be 2°C or 3°C higher.

Buildings break up the wind and can reduce wind speeds by up to a third. Sometimes the wind can increase speed as it rushes around buildings.

Aspect

The direction in which a place is facing is called its aspect. Places facing the sun will be warmer than those in shadow.

In Britain the sun rises in the east and moves through the south before it sets in the west. South-facing places get most of the sun and are usually the warmest.

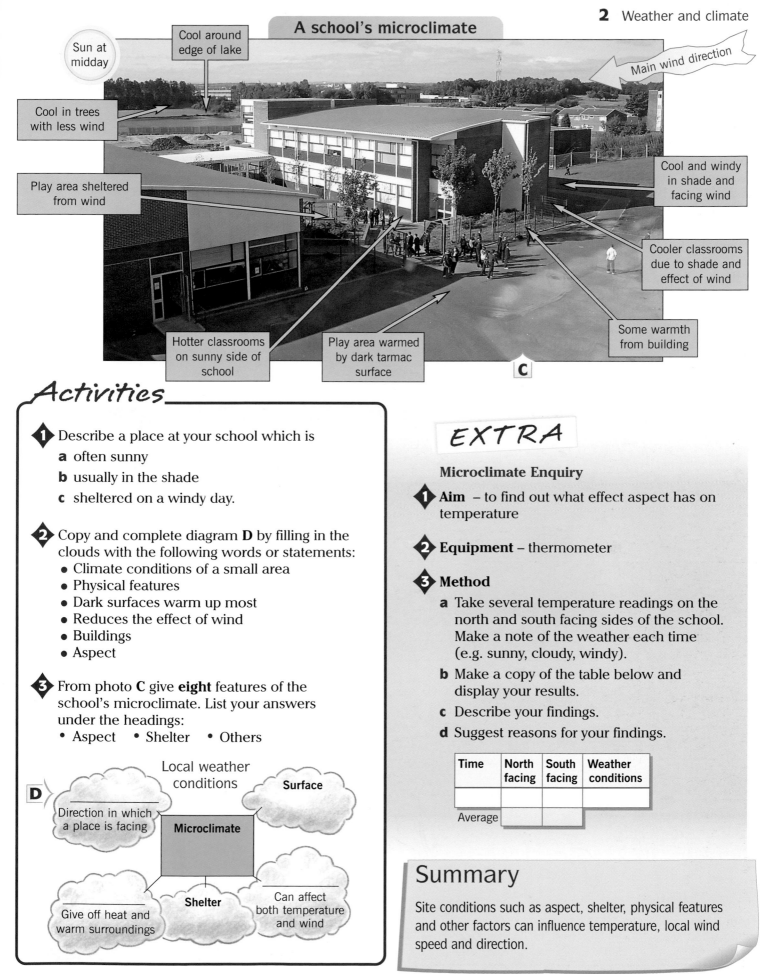

A school's microclimate

Sun at midday

Cool around edge of lake

Main wind direction

Cool in trees with less wind

Play area sheltered from wind

Cool and windy in shade and facing wind

Cooler classrooms due to shade and effect of wind

Some warmth from building

Hotter classrooms on sunny side of school

Play area warmed by dark tarmac surface

C

Activities

1 Describe a place at your school which is
  a often sunny
  b usually in the shade
  c sheltered on a windy day.

2 Copy and complete diagram **D** by filling in the clouds with the following words or statements:
  • Climate conditions of a small area
  • Physical features
  • Dark surfaces warm up most
  • Reduces the effect of wind
  • Buildings
  • Aspect

3 From photo **C** give **eight** features of the school's microclimate. List your answers under the headings:
  • Aspect • Shelter • Others

D

Local weather conditions

Surface

Direction in which a place is facing

Microclimate

Give off heat and warm surroundings

Shelter

Can affect both temperature and wind

EXTRA

Microclimate Enquiry

1 Aim – to find out what effect aspect has on temperature

2 Equipment – thermometer

3 Method
  a Take several temperature readings on the north and south facing sides of the school. Make a note of the weather each time (e.g. sunny, cloudy, windy).
  b Make a copy of the table below and display your results.
  c Describe your findings.
  d Suggest reasons for your findings.

Time	North facing	South facing	Weather conditions
Average			

Summary

Site conditions such as aspect, shelter, physical features and other factors can influence temperature, local wind speed and direction.

What is Britain's weather?

Weather is what happens in the atmosphere day by day but **climate** is different. It is the weather taken on average over many years. Climate is about warm dry summers, cool wet winters or, as at the North and South Poles, being cold all year. In Britain the weather is always a popular topic of conversation probably because it is always changing or it's never quite what we want it to be. Changes also occur in the climate. It can change from time to time (seasonal) or it may be different from place to place.

Temperature

The average monthly temperatures for summer and winter are shown on maps **A** and **B**. If you look closely you should see three main differences. These differences are explained below.

1 As expected, temperatures are higher in summer than in winter.

2 Temperatures at any one time are not the same all over Britain.

3 The pattern of temperature is different in the two seasons.

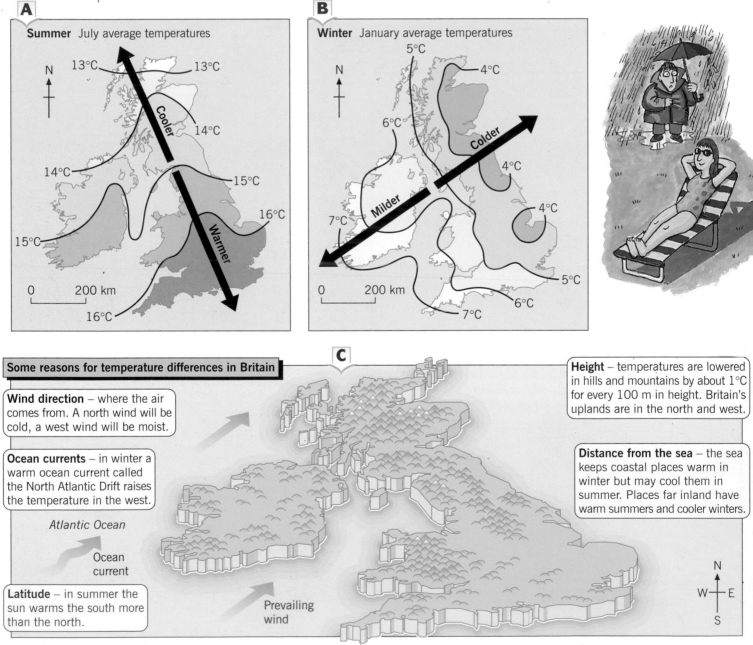

A Summer July average temperatures

13°C 13°C Cooler 14°C 14°C 15°C 15°C 16°C Warmer 16°C 0 200 km

B Winter January average temperatures

5°C 4°C 6°C Colder 4°C 7°C Milder 4°C 5°C 6°C 7°C 0 200 km

Some reasons for temperature differences in Britain

Wind direction – where the air comes from. A north wind will be cold, a west wind will be moist.

Ocean currents – in winter a warm ocean current called the North Atlantic Drift raises the temperature in the west.

Atlantic Ocean

Ocean current

Latitude – in summer the sun warms the south more than the north.

Prevailing wind

Height – temperatures are lowered in hills and mountains by about 1°C for every 100 m in height. Britain's uplands are in the north and west.

Distance from the sea – the sea keeps coastal places warm in winter but may cool them in summer. Places far inland have warm summers and cooler winters.

N
W E
S

Rainfall

In Britain we can expect rain at any time of the year. Although winter is wetter than the summer, seasonal differences in rainfall are very small.

As map **D** shows, however, the amount of rainfall varies considerably from place to place and the greatest differences are between the east and the west.

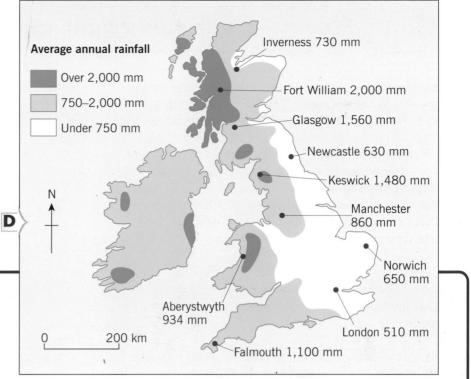

D

Average annual rainfall

- Over 2,000 mm
- 750–2,000 mm
- Under 750 mm

Inverness 730 mm
Fort William 2,000 mm
Glasgow 1,560 mm
Newcastle 630 mm
Keswick 1,480 mm
Manchester 860 mm
Norwich 650 mm
Aberystwyth 934 mm
London 510 mm
Falmouth 1,100 mm

N

0   200 km

Activities

1 What is the difference between weather and climate?

2 a Write out and complete the following sentence to describe summer temperatures in Britain.

> Summers in Britain are _____ than winter. The warmest weather is in the _____ and temperatures get lower (decrease) towards the _____ .

b Write a similar sentence to describe winter temperatures.

3 Why are there temperature differences in Britain? Think of **three** reasons and write them in your workbook.

4 a List the **three** wettest and the **three** driest towns from map **D**. Give your answers in order with the wettest first.

b With the help of a simple diagram, describe the difference in rainfall from east to west. Give actual figures in your answer.

5 a Make a large copy of map **E**.

b Match the following climate descriptions to Ⓐ, Ⓑ, Ⓒ and Ⓓ and write them on your map.
Ⓐ has been done on the map to help you.

- Warm summers, cold winters, dry
- Mild summers, mild winters, wet
- Warm summers, mild winters, quite wet
- Mild summers, cold winters, dry

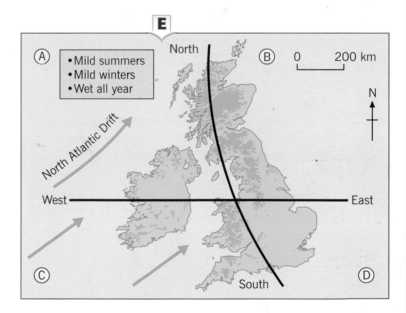

E

Ⓐ
- Mild summers
- Mild winters
- Wet all year

North Atlantic Drift

North

West ——————— East

South

Ⓑ   0   200 km

N

Ⓒ   Ⓓ

c Suggest reasons for the climate of area Ⓐ.

d Mark where you live on your copy of map **E**.

e Describe the climate there and suggest reasons for it.

Summary

Britain's climate varies from place to place and from season to season. Heating from the sun, ocean currents, and the height of the land are some of the reasons for these variations.

How does it rain?

The Atacama Desert in South America has had no rain for over 400 years yet parts of the Amazon rainforest, also in South America, have rain on more than 330 days each year. Seathwaite in the Lake District, the wettest place in England, has on average 3,340 mm of rain per year, whilst Newcastle, only 130 km away, may expect just 630 mm.

What are the reasons for this, what causes rain, and why are some places wetter than others?

Clouds are made up of extremely tiny drops of moisture called **cloud droplets**. They are only visible because there are billions of them crowded together in a cloud.

Clouds form when moist air rises, cools and changes into cloud droplets. This is **condensation**. A cloud gives rain after these tiny cloud droplets grow thousands of times larger into raindrops which then fall to the ground.

Look at diagram **A**. It shows how rain is formed. The process is always the same: air rises, cools, condenses and precipitates.

Air can be forced to rise in three different ways. This gives the three main types of rainfall: **relief**, **convectional** and **frontal**. These are shown in diagrams **B**, **C** and **D**.

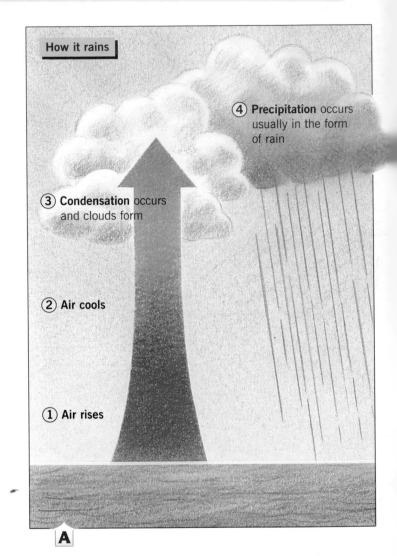

How it rains

④ **Precipitation** occurs usually in the form of rain

③ **Condensation** occurs and clouds form

② **Air cools**

① **Air rises**

A

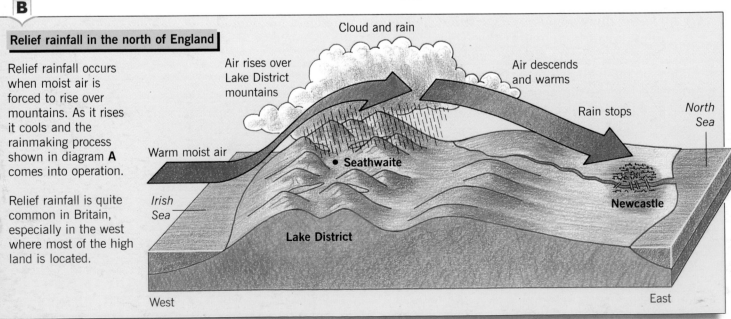

B

Relief rainfall in the north of England

Relief rainfall occurs when moist air is forced to rise over mountains. As it rises it cools and the rainmaking process shown in diagram **A** comes into operation.

Relief rainfall is quite common in Britain, especially in the west where most of the high land is located.

Cloud and rain

Air rises over Lake District mountains

Air descends and warms

Rain stops

North Sea

Warm moist air

Irish Sea

• Seathwaite

Newcastle

Lake District

West

East

Convectional rainfall

Rising air cools

Clouds and rain

Warm air rises

Ground warmed by sun

When the ground surface is heated by the sun, the air above is warmed up. This air rises and as it cools down clouds form and rain follows. The showery weather and thunderstorms of a British summer are this type of rainfall.

C

Frontal rainfall

Clouds and rain

Warmer, lighter air rises over heavier, colder air

Rising air cools

Warm air

Cold air

When a mass of warm air meets air at a lower temperature, it rises up and over the colder, heavier air. Once it is made to rise, cloud and rain will follow due to the process shown in diagram **A**.

The place where warm air and cold air meet is called a **front**. Frontal rainfall is very common in Britain throughout the year and especially in winter.

D

Activities

1 Match the beginnings of the labels in **E** to their correct endings

2 With the help of a labelled diagram, describe how it rains.

3 a Make larger copies of the three diagrams in **F**.

b For each diagram explain how it rains by adding labels at points ①, ②, ③ and ④.

c Add colour to make your diagrams clearer.

d Underneath each of your diagrams give a brief reason for the air rising.

e Give each diagram a title.

E

Clouds are — rain, snow and other forms of moisture in the sky.

Precipitation is — when water vapour changes to water.

Condensation happens — made up of tiny drops of moisture called cloud droplets.

4 Explain why Seathwaite is wetter than Newcastle. Use diagram **B** to help you.

F Mountain / Sun's heat / Warm / Cold

Summary

Rain is caused by moist air rising and cooling. The three types of rainfall produced in this way are relief, convectional and frontal.

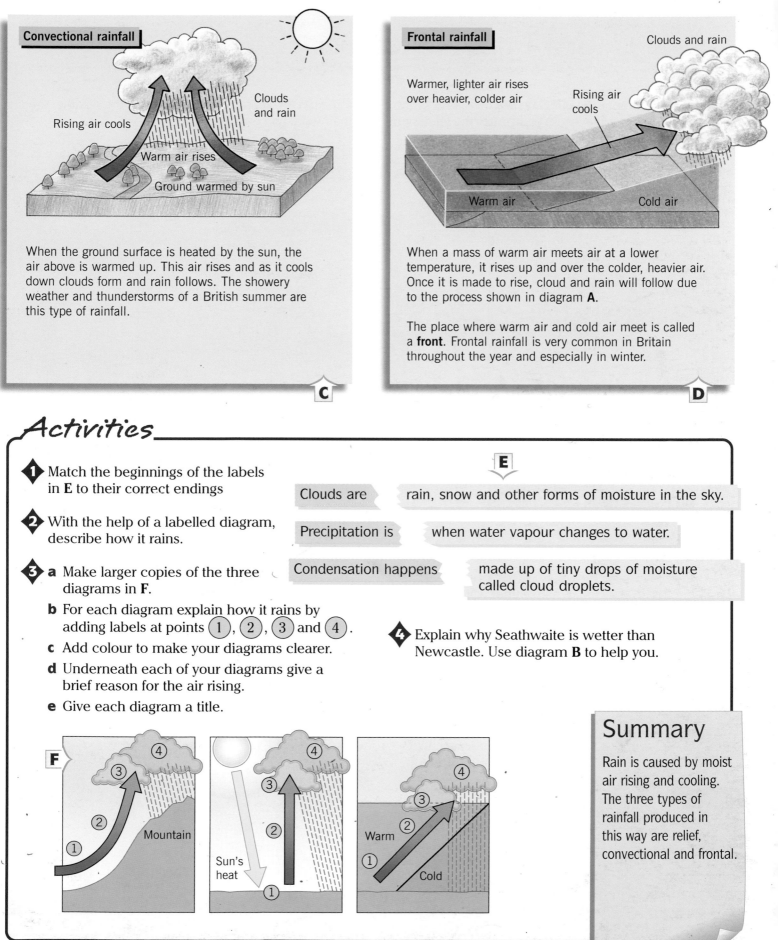

Forecasting the weather – anticyclones

Weather has an important effect on our lives. Every day in the newspapers and every evening after the television news there is a **weather forecast**. Forecasts can tell us in advance what the weather will be. For many of us they are of passing interest but for some people such as farmers, fishermen, aircraft pilots and builders the forecasts are very important because the weather affects their work and even their safety.

Map **A** is a typical newspaper weather map. Notice how easy it is to read the weather using the picture symbols.

Forecast for noon 30 June

A

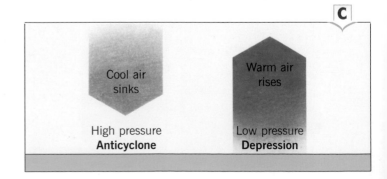

B Satellite photo of an anticyclone

How do weather forecasters know what the weather will be like tomorrow? How can they tell if it will be wet or dry, or hot or cold?

Forecasting is very complicated and lots of information and advanced computers are needed to make good forecasts. In recent times, satellites have become particularly useful because they can see weather systems many kilometres away.

Photo **B** was taken from a satellite. It shows Britain with very little cloud overhead and clearly enjoying a fine sunny day. Photos like these are taken every few hours and by looking back over several of them the movements of the weather systems can be worked out, and forecasts made.

The weather system in photo **B** is an **anticyclone**. It occurs because of changes in the air pressure. The weight of air pressing down on us from above is called pressure. This pressure varies from place to place and results in the development of pressure systems. Areas with above average pressure (high pressure) are called anticyclones and usually give good weather. Areas with less than average pressure (low pressure) are called depressions and usually give poor weather.

C

Cool air sinks

Warm air rises

High pressure
Anticyclone

Low pressure
Depression

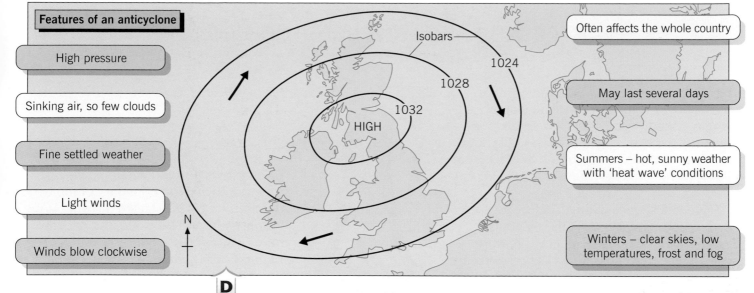

Features of an anticyclone

High pressure

Sinking air, so few clouds

Fine settled weather

Light winds

Winds blow clockwise

Often affects the whole country

May last several days

Summers – hot, sunny weather with 'heat wave' conditions

Winters – clear skies, low temperatures, frost and fog

Isobars

1024

1028

1032

HIGH

N

D

Activities

1 From map **A**, give the weather that is forecast for the place where you live.

2 a When do you think it would be useful for you to know the next day's weather?

b Make a list of people who need the weather forecast. For each person explain why they need to know about the weather.

3 How do satellites help in forecasting weather conditions?

4 a Make a sketch of an anticyclone like the one in diagram **D** above.

b Next to your sketch, write out the paragraph below and fill in the blank spaces with the following words:

• LONG • LARGE • HIGH • COOL

Anticyclones are areas of _____ pressure which form when _____ air sinks. They usually cover _____ areas and give _____ periods of fine settled weather.

5 Copy and fill in table **F** to show the weather features of an anticyclone.

E Weather in a winter anticyclone

6 Study map **A** on page 32 giving the newspaper weather forecast. Write a weather forecast to be read out on the radio for the same day. Your forecast should be about 100 to 150 words in length.

F

	Summer	Winter
Temperatures		
Cloud cover		
Wind speed		
Wind direction		
Rain		
Other features		

Summary

Knowing what the weather will be like can be useful to us. Anticyclones can bring good weather and may be forecast with the help of satellites.

Forecasting the weather – depressions

All too often we seem to hear the weather forecast begin with 'Today will be cloudy, and rain already in the west will spread eastwards to cover all areas by late afternoon.' The reason for this is that for much of the year Britain is affected by low pressure.

As diagram **A** shows, at times of low pressure the air is usually rising. As it rises it cools, condenses and clouds form. Low pressure areas are called **depressions**. Depressions are the most important weather systems affecting Britain and they bring with them clouds and rain.

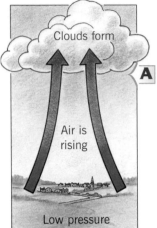

Clouds form

Air is rising

Low pressure

A

Depressions develop where warm air meets cold air. The boundary of the two different air types is called a **front**. Along a front there will be cloud and usually rain. Diagram **B** shows the features of a depression. The **isobars** are lines that join up areas of equal pressure and they help us to see the shape of the depression.

B

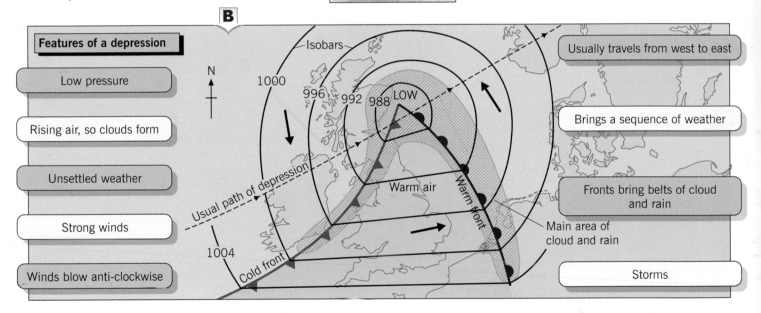

Features of a depression

- Low pressure
- Rising air, so clouds form
- Unsettled weather
- Strong winds
- Winds blow anti-clockwise

Isobars

N

1000
996 992 988 LOW

Usual path of depression

Warm air

Warm front

Cold front

1004

Usually travels from west to east

Brings a sequence of weather

Fronts bring belts of cloud and rain

Main area of cloud and rain

Storms

C Satellite photo of a depression

Depressions are huge areas of low pressure measuring many hundreds of kilometres across. They show up very clearly on satellite photographs as great swirls of cloud that look like gigantic catherine wheel fireworks. The fronts are easily recognised as areas of thick white cloud arranged in an upside down 'V' shape. The centre of the depression is normally just above or a little behind the point of the 'V'.

Look at photo **C** which shows a depression approaching Britain. Can you work out which areas are the fronts and where the centre of the depression might be? With help from diagram **B** can you work out which is the area of warm air? What sort of weather does that area seem to have?

Depressions usually form over the Atlantic Ocean and move across Britain from west to east. With help from satellite photographs, weather forecasters can work out the direction they are travelling and how fast they are moving. From this information they can produce quite accurate weather forecasts. Diagram **D** shows how the weather changes as a depression passes over Britain. Notice the changes in the weather that occur in the area where you live.

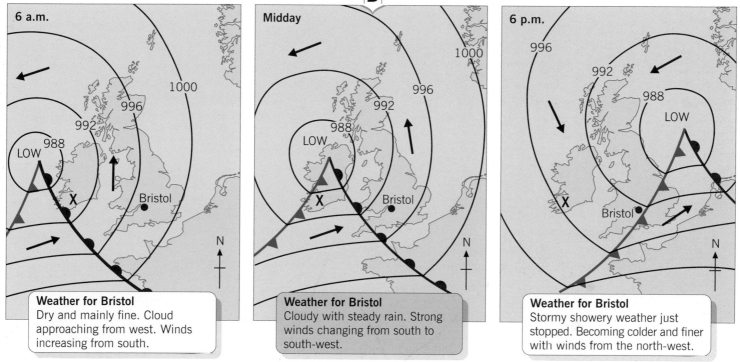

D A depression passing over Britain

Weather for Bristol
Dry and mainly fine. Cloud approaching from west. Winds increasing from south.

Weather for Bristol
Cloudy with steady rain. Strong winds changing from south to south-west.

Weather for Bristol
Stormy showery weather just stopped. Becoming colder and finer with winds from the north-west.

Activities

1 The words below have been jumbled up. Unscramble the words and fill in the blank spaces in the following paragraph.

> NIRA   SATE   DULOC   OWL   TEWS   SERIS
> Depressions are areas of _____ pressure which form when air _____ . They usually move across Britain from _____ to _____ and bring most of our _____ and _____ .

2 With the help of a labelled diagram, explain why depressions bring cloud and rain.

3 Make a labelled sketch of a depression like the one shown in diagram **B**. Underneath your sketch make a copy of the table below.

Complete the table to show the main features of a depression.

General features	Weather

4 From diagram **D**:

a Describe the weather at place **X** for 6 a.m., 12 midday and 6 p.m.

b Explain why the weather has changed.

c At what time will the warm front be over the place where you live?

d Describe the weather you may get at that time.

5 a Trace the outline of Britain from photo **C**.

b Mark and label the following:
- warm front
- cold front
- warm air
- centre of the depression.

c Shade the area of cloud and rain along the fronts.

d Describe the weather over Britain.

Summary

Depressions are the most common weather system affecting Britain. They are low pressure areas and bring stormy winds, cloud and rain.

This enquiry is concerned with the weather and climate. Pages 28 and 29 of this book may be helpful to you as you work through the enquiry. Your task is to reply to a letter sent to you by a company in America. To do this you will need to look closely at Britain's weather and climate and answer the enquiry question given below.

There should be three main parts to your enquiry.

◆ The first part will be an introduction. Here you can explain what the enquiry is about.

◆ In the next part you will need to collect and present information about Britain's weather and climate. You can then use that information to answer the question set.

◆ Finally you will need a conclusion. Here you could write a letter to explain your findings.

What are the differences in weather and climate across Britain?

World Wide Leisure Corporation

174 Aspen Boulevard,
Denver
Colorado 96541, USA

Tel/Fax (303)569-3309

Dear Sir/Madam

I am the Personnel Manager for a large American company. We are planning to open four offices in Britain. These offices will be at Oban, Aviemore, Plymouth, and Cambridge. Each manager will bring the family with them, and they are likely to stay for three years.

Like many Americans, each family is keen on leisure and doing things out of doors. Of course these activities in turn depend upon the weather and climate. Each family has different interests – as will be listed later in this letter. We are therefore allowing each of them to choose the office in the region where the weather and climate best suits their interests.

To help them do this we would appreciate your help. Please could you give us the weather and climate for the four places and suggest which you think is best suited to each of the families.

Yours sincerely

John F. Gates

John F. Gates
Personnel Manager

1 Introduction – what is the enquiry about?
You will need to use maps and writing here. Star diagrams or lists might also help.

a First look carefully at the enquiry question above and say what you are going to try to find out.
- Give the meanings of **weather** and **climate**. Pages 24, 28 and the Glossary will help you.
- Describe how Britain's weather and climate can be roughly divided into four regions.
- Explain where these regions are, and then describe the different conditions in them.

b Describe briefly what the letter has asked you to do. Show on a map where the four places are located. List the particular features of weather and climate that you will look at.

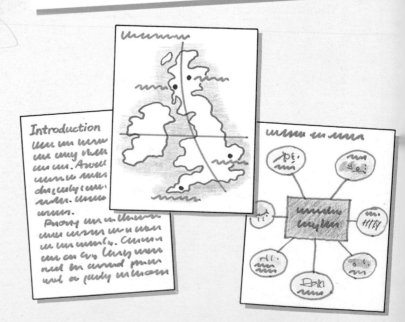

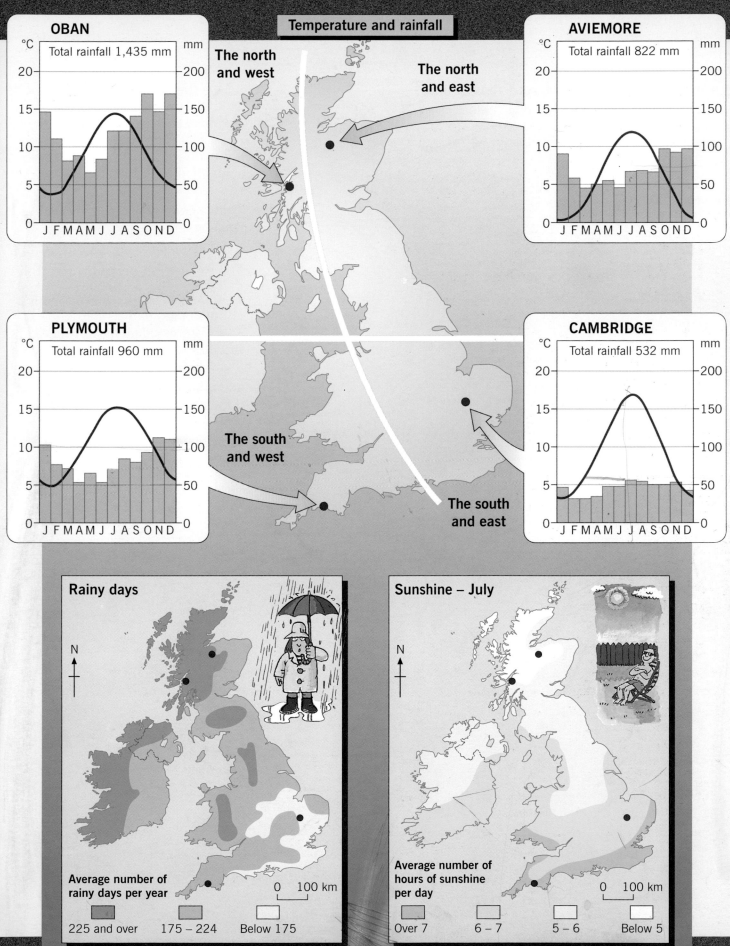

Temperature and rainfall

OBAN

Total rainfall 1,435 mm

The north and west

The north and east

AVIEMORE

Total rainfall 822 mm

PLYMOUTH

Total rainfall 960 mm

The south and west

CAMBRIDGE

Total rainfall 532 mm

The south and east

Rainy days

N

Average number of rainy days per year

0   100 km

225 and over   175 – 224   Below 175

Sunshine – July

N

Average number of hours of sunshine per day

0   100 km

Over 7   6 – 7   5 – 6   Below 5

2 **What is Britain's weather and climate like?**

a Make a large copy of the table below of Britain's weather.

b Complete the table using information from this page and from page 37.

Britain's weather	Oban (north and west)	Aviemore (north and east)	Plymouth (south and west)	Cambridge (south and east)
January temperature (°C)				
July temperature (°C)				
January rainfall (mm)				
July rainfall (mm)				
Total rainfall (mm per year)				
Rainy days (number per year)				
July sunshine (hours per day)				
Snow lying (days per year)				
Average wind strength (description and km/h)				

3 **Where is the best weather?**

Each family made a list of the weather and climate that they would like to have for their stay in Britain. This information is given on page 39. You can now find out which places are most suited to each family.

a Make a copy of the table for the Jackson family.

b For each place in turn put:
  ✔ a tick if the weather is suitable
  ✗ a cross if it is unsuitable
  ? a question mark if it is not perfect but not too bad.

Your completed table showing Britain's weather will give you all the answers for this.

c Add up the ticks to find which place is the most suitable. The one with the most ticks would be the best.

d Now repeat the parts **a**, **b** and **c** for each of the other three families.

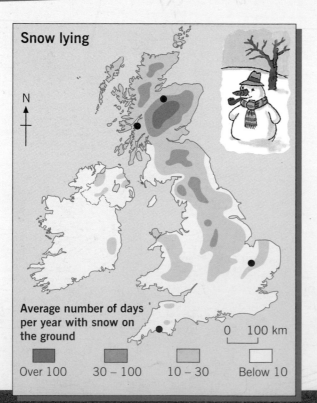

Snow lying

N

Average number of days per year with snow on the ground

0   100 km

| Over 100 | 30 – 100 | 10 – 30 | Below 10 |

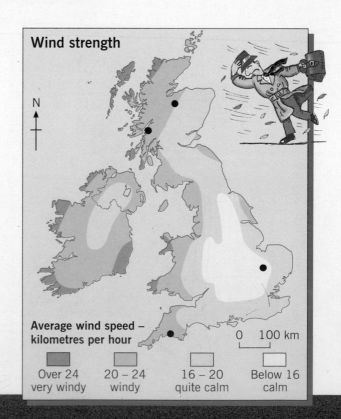

Wind strength

N

Average wind speed – kilometres per hour

0   100 km

| Over 24 very windy | 20 – 24 windy | 16 – 20 quite calm | Below 16 calm |

> We are a cycling family so we don't like rain or wind. We prefer warm summers and cold winters.

> We prefer it not to be cold or too snowy. We like to go fishing so rainy days can be good for us.

The Jackson family	Oban (north and west)	Aviemore (north and east)	Plymouth (south and west)	Cambridge (south and east)
Cold winters (Jan. temp. below 3°C)				
Warm summers (July temp. 15–20°C)				
Dry (less than 175 rainy days)				
Quite sunny in summer (6–7 hrs per day)				
Very little wind (below 16 km/h)				
TOTAL				

The Houston family	Oban (north and west)	Aviemore (north and east)	Plymouth (south and west)	Cambridge (south and east)
Mild winters (Jan. temp. 3–7°C)				
Mild summers (July temp. 10–14°C)				
Many rainy days (over 225 per year)				
A little snow (10–13 days per year)				
Windy (20–24 km/h)				
TOTAL				

> We are keen walkers and skiers. Our favourite days are in winter when it is cold and snowy.

> We enjoy barbecues and relaxing in the sun. We like warm sunny summers. Rain doesn't bother us but we really don't like the cold.

The Grant family	Oban (north and west)	Aviemore (north and east)	Plymouth (south and west)	Cambridge (south and east)
Cold winters (Jan. temp. below 3°C)				
Mild summers (July temp. 10–14°C)				
Quite dry (total rain 600–900 mm)				
Cloudy summers (under 5 hrs per day)				
Lots of snow (over 30 days per year)				
TOTAL				

The Stolberg family	Oban (north and west)	Aviemore (north and east)	Plymouth (south and west)	Cambridge (south and east)
Mild winters (Jan. temp. 3–7°C)				
Warm summers (July temp. 15–20°C)				
Quite wet (total rain 900–1,200 mm)				
Lots of summer sunshine (over 7 hrs per day)				
Windy (20–24 km/h)				
TOTAL				

4 Conclusion

Now you should look carefully at your work and answer the enquiry question. You could do this by replying to the letter from the World Wide Leisure Corporation. This could include writing and perhaps a labelled map.

a Describe the weather and climate in each of the four regions of Britain. Your completed tables from this page will help you.

b Say which place is best suited to each of the four families. Give reasons for your answer.

Why is flooding a problem?

What is this unit about?

This unit is about the causes and effects of river flooding. It shows how flooding can affect people in different ways and suggests how the risk of future flooding may be reduced.

In this unit you will learn about:

◆ the water cycle

◆ what happens to rain when it reaches the ground

◆ the causes of flooding and how individuals and communities respond to the problem

◆ how the effects of flooding in Britain are different from those in Bangladesh

◆ how the risk of flooding can be reduced.

A Bangladesh

B Bangladesh

40

Why is this flooding topic important?

Flooding is an increasing problem across the world. More than 7 million people in the UK are now at risk from flooding every year. Even if flooding has not affected you yet, it could easily do so some time in the future. For these reasons, we need to understand the causes and effects of flooding so that we can try to manage the problems they create.

This unit can also help you to:
- be aware of the effects of flooding
- find out if flooding is a problem where you live
- prepare for a flooding situation
- know what to do during and after a flood
- be able to help other people affected by flooding.

C | Carlisle

D | Carlisle

- Look at the people in the photos.
  - What problems do they face?
  - What help do they need?
  - Who needs the most help?
  - How could they prepare for the situation?
  - What could you do to help them?

STRUTS
...TY SUPERSTORE

OPEN SUNDAYS

HOUSE CLEARANCE

FACE PAINTS

HM COASTGUARD

How does the water cycle work?

Rainy days are annoying and a nuisance to most of us. Yet rain is very important to our world as it is part of a never-ending cycle in which water is used over and over again. This cycle is called the **water cycle**. The amount of water in the cycle always stays the same. Some of the water may be **stored** in the sea, in the air or on land. Later, some of this water will be moved or **transferred** around the cycle. The main **stores** and **transfers** in the cycle are shown in diagram **A**.

The water cycle can be very complicated but its main features are shown in diagram **B**. Notice that the water can be moved in different forms – as vapour, rain, snow or hail. Some of the geographical terms used on this diagram are long and will be new to you. Chart **C** on the next page explains what these words mean.

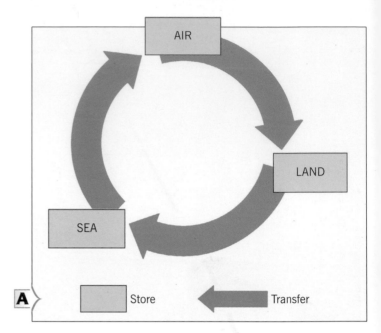

A

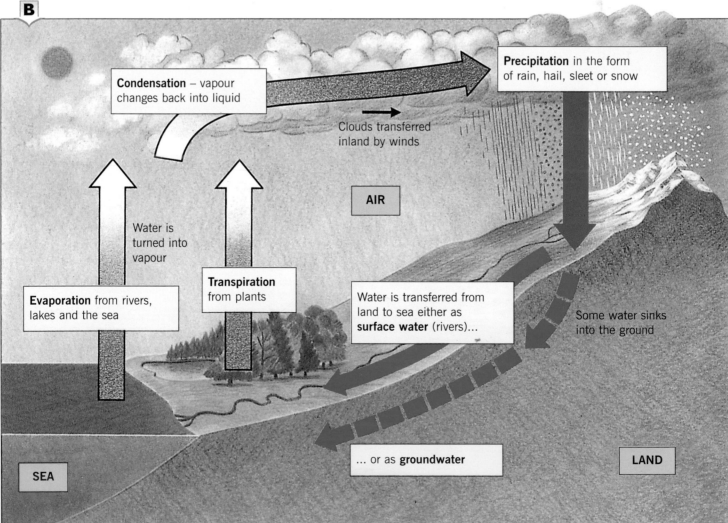

B

Evaporation	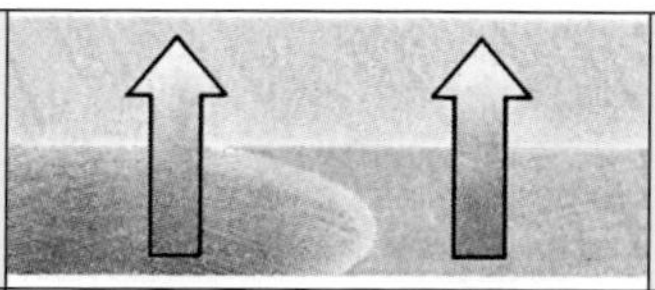	The transfer and change of water from the ground into water vapour in the air. Water vapour is an invisible gas.
Transpiration		The transfer and change of water from plants into water vapour in the air.
Condensation		Water vapour in the air changes back into a liquid. It forms small droplets which are visible as cloud.
Precipitation	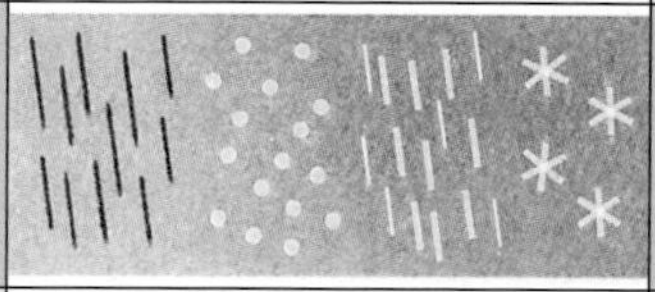	The transfer of water from the air to the land. Water can fall to earth as rain, hail, sleet or snow.
Surface water	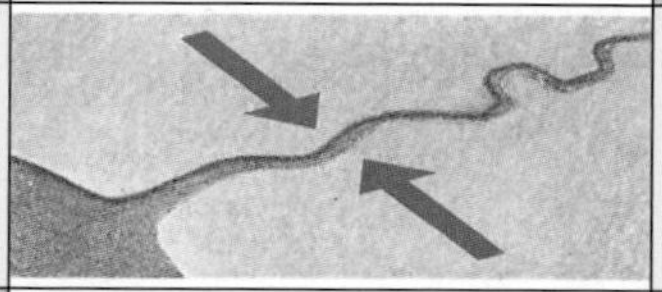	The transfer of water back to the sea over the ground surface. It is called surface run-off. It is easiest to see where it forms rivers.
Groundwater		The transfer of water through the ground back to the sea.

C

Activities

1 Diagram **D** shows part of the water cycle. Draw the diagram and complete the boxes. Choose your answers from the following:

condensation   evaporation

groundwater   precipitation

surface water   transpiration

2 What would happen to surface water (the river) if there was:

a an increase in rainfall

b a decrease in rainfall

c a lot of snow which did not melt

d a lot of snow which melted very quickly?

D

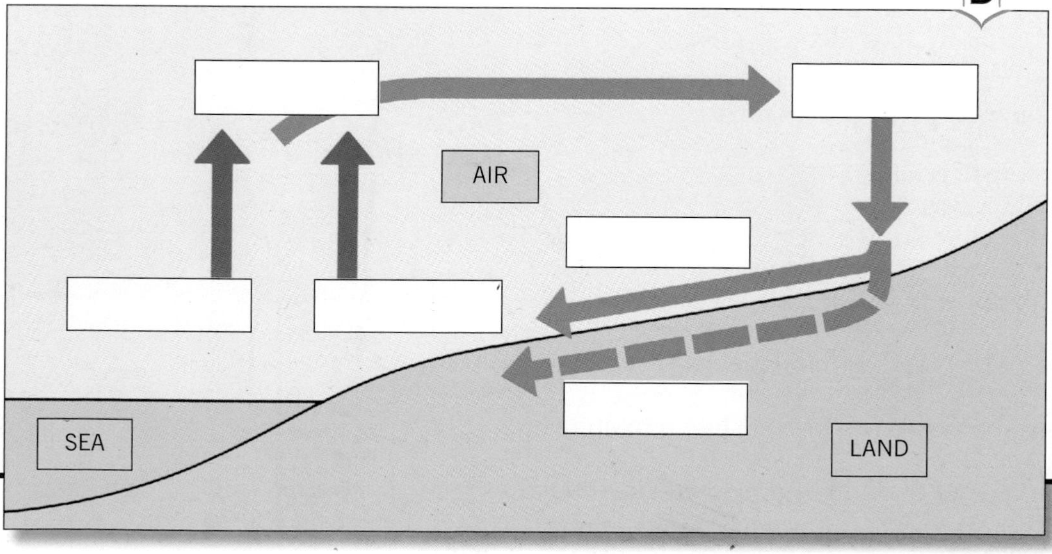

AIR

SEA

LAND

Summary

Water can be stored in the sea, in the air and on land. The water cycle is the never-ending transfer of this water between the sea, the air and the land.

43

What is a river basin?

Most water that falls on land as rain eventually finds its way into a river. This is an important part of the **water cycle** which was explained on pages 42 and 43. Rain is a type of **transfer** whilst the river is a **store**. Although rivers differ in many ways, they all have similar features. Some of these are shown in **B** below.

If you look at the top of a large tree you will see lots of twigs. Twigs are small branches. If you follow these downwards, you will see these twigs joining to form branches. These branches in turn join to form one big trunk.

A **river** is like a tree. It has lots of small streams which join to form **tributaries** which later join to form the main river. When it rains, most of the water slowly drains into streams, then into tributaries and finally into the main river. A **river basin** is the area of land drained by a main river and its tributaries.

A

Twigs

Main trunk

B

A **river basin** is an area of land where rain collects. The river basin of the Amazon is the size of Europe.

A river begins at its **source**. The source of the Amazon is 6,500 km from the sea.

A **tributary** is a small river. Tributaries flow into a main river.

The boundary or edge of a river basin is called a **watershed**. It is usually on high ground.

Rivers flow in a **channel** (photo **C**). The channel has banks and a bed. Floods occur when a river overflows its channel.

Rivers flow into the sea or a lake. The end of a river is called the **mouth**. The mouth of the Amazon is 50 km wide.

C River Amazon

Activities

Diagram **B** shows a river basin. There are two lists below. One gives words used to describe parts of river basins and the second gives their meanings. Match up the two lists.

A watershed	is where a river begins
The source	is where the river flows
A river basin	is where the river flows into a lake or the sea
A tributary	is an area of highland forming the edge of a river basin
A channel	is a stream or small river flowing into a main river
The mouth	is an area of land drained by a river and its tributaries

Where are the world's most important rivers?

Activities

Thirteen important world rivers are, in alphabetical order, the Amazon, Colorado, Danube, Ganges, Murray-Darling, Mississippi, Nile, Rhine, St Lawrence, Volga, Yangtze, Zambezi and Zaire (or Congo).

1 Fit the names of these rivers into crossword **D** on the right. The number of letters in each word will help you to fit them into the puzzle. For example, the Murray-Darling, which has been done for you, was the only word with 13 letters. It could only fit into that one place.

2 With help from map **E** below, sort the rivers into groups under the headings: *Africa, America, Asia, Australia* and *Europe*.

3 The three longest rivers in the United Kingdom are the Severn, the Thames and the Trent. Name **one** city on each of these rivers. The map on the back cover will help you.

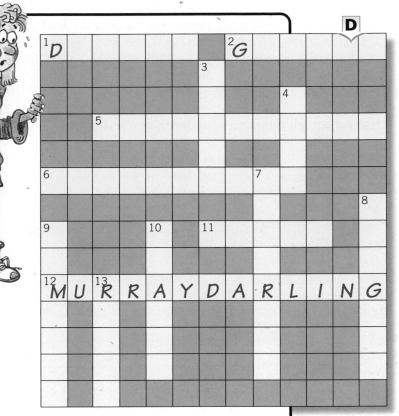

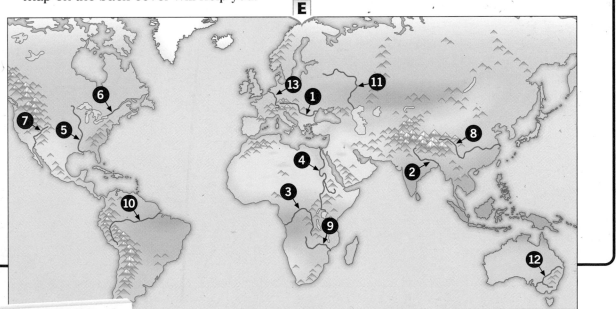

E

EXTRA

Match the 13 rivers named in activity **1** above with a country through which they flow. Some flow through several countries, so choose only **one** important country. An atlas will help you with this.

Summary

Rain collects in rivers in a river basin. Rivers have their source in highland areas and flow in a channel to the sea or to a lake.

What causes a river to flood?

All of the water that flows down a river comes from rain or melting snow. Sometimes after heavy rain or a rapid snow melt, there may be too much water for the river to hold. The river will then overflow its banks and spread out across the land on either side of its channel. This is called a **river flood**.

Usually when it rains, most water simply soaks into the ground and there is little chance of a flood. If, however, the water is unable to soak into the ground, it will stay on the surface and flow quickly downhill and into the river. This is when floods are most common.

Some rivers are more at risk from flooding than others. Put simply, heavy rain and anything which stops that rain from soaking into the ground will increase the chances of flooding. Some of the factors that increase the risk of flooding are shown below.

A

The rapid melting of snow can cause flooding

B

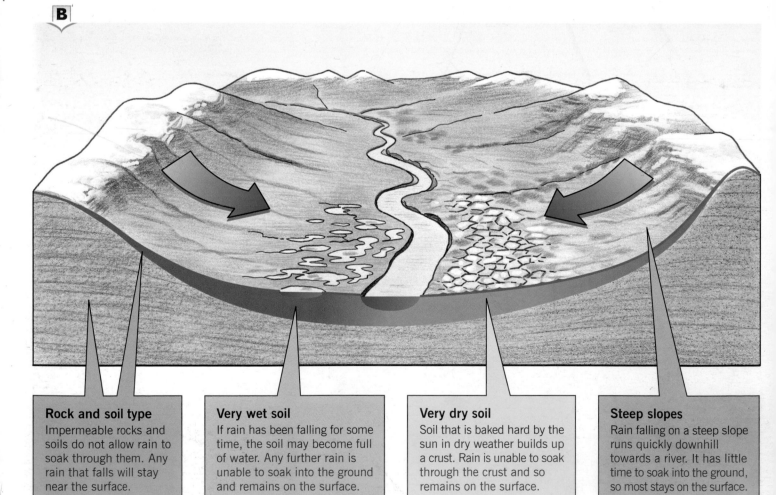

Rock and soil type
Impermeable rocks and soils do not allow rain to soak through them. Any rain that falls will stay near the surface.

Very wet soil
If rain has been falling for some time, the soil may become full of water. Any further rain is unable to soak into the ground and remains on the surface.

Very dry soil
Soil that is baked hard by the sun in dry weather builds up a crust. Rain is unable to soak through the crust and so remains on the surface.

Steep slopes
Rain falling on a steep slope runs quickly downhill towards a river. It has little time to soak into the ground, so most stays on the surface.

Floods are more common now than they used to be. There are more of them and they are increasing in size. Many people are blaming human activity for this.

Two ways in which humans may increase the risk of flooding are by cutting down trees and building more towns and cities. These are shown in drawings **C** and **D**.

C

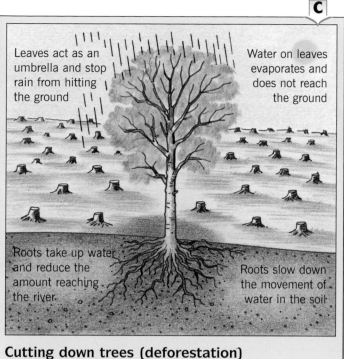

Leaves act as an umbrella and stop rain from hitting the ground

Water on leaves evaporates and does not reach the ground

Roots take up water and reduce the amount reaching the river

Roots slow down the movement of water in the soil

Cutting down trees (deforestation)

Many of the world's forests are being cleared to make way for other developments. In some countries the number of serious floods has more than doubled since large-scale tree clearing began.

D

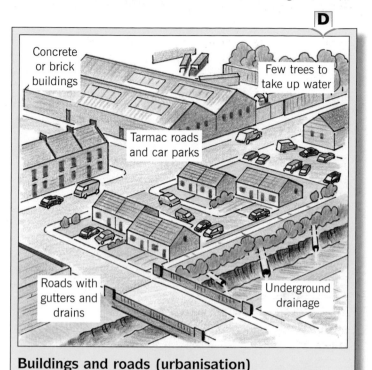

Concrete or brick buildings

Few trees to take up water

Tarmac roads and car parks

Roads with gutters and drains

Underground drainage

Buildings and roads (urbanisation)

Rain falling on concrete and tarmac is unable to soak into the ground, so stays on the surface. Gutters and drains then carry the water quickly and directly to the river. Large towns are most at risk.

Activities

1
  a Make a larger copy of drawing **E**.

  b Add the following labels to your drawing to show how a river floods:
- River level rises
- Water quickly reaches river
- River floods
- Water runs over surface
- Heavy rain falls
- Rain soaks into ground

E

2 Describe four factors that increase the risk of flooding.

3 With the help of diagram **F**, describe how:
  a cutting down trees, and
  b building towns can make floods worse.

F

Summary

River flooding is most likely after heavy rain or rapid snow melt. The flood risk is greatest when water is unable to soak into the ground. Human activities can increase the chance of flooding.

Floods in the UK, 2000

Evening COURIER

Sussex
Friday, 13 October 2000

FLOOD HAVOC HITS SOUTHERN ENGLAND

Amassive rescue operation was underway last night as tens of thousands of homes were hit by the UK's worst floods for 30 years.

More than a month's rain fell in 24 hours as torrential downpours and storms caused six of this region's rivers to burst their banks. Thousands of homes, shops and offices filled with water up to 2 metres deep. Most of the main roads in the region were blocked and all rail services cancelled.

Emergency services, including lifeboats, coastguard helicopters, the fire service and police were scrambled to rescue hundreds of people trapped in buildings and on rooftops.

Worst hit was Uckfield which was completely submerged and cut off after a wall of water, almost 2 metres in height, crashed through the town centre. Cars were swept away and the soggy contents of shops floated into the streets. Thousands of homes in the area suffered severe damage. Many people have lost all of their belongings. The council have set up refuge centres at a local school and leisure centre.

As the flood water subsides, people returning home are finding everything covered in a thick layer of foul-smelling mud. The clean-up operation will take weeks. Many people will not be back in their homes for months and some businesses may never re-open.

Insurance experts are putting the cost at over £500 million but there are fears that many people will not be insured. The government has promised help to areas most in need.

The south of England has had many floods recently. Most have been caused by very heavy rain. Floods like these are called **flash floods** because they happen very suddenly and last for just a short time. Flash floods can be dangerous as they come without warning and give people little time to escape.

A serious flood rarely has one single cause, though. In the south of England much of the land is low-lying and many new developments have been built on the floodplain. Some river embankments collapsed, and reservoirs that had reached the point of overflowing added to the problem.

The October 2000 flood

1 Heavy rain had been falling in the area for more than a week.

2 The ground became full of water and could take no more.

3 Almost two months of rain fell on the area in less than 24 hours.

4 Rivers burst their banks and water flooded surrounding areas.

5 Recent building in floodplain areas made the problem worse.

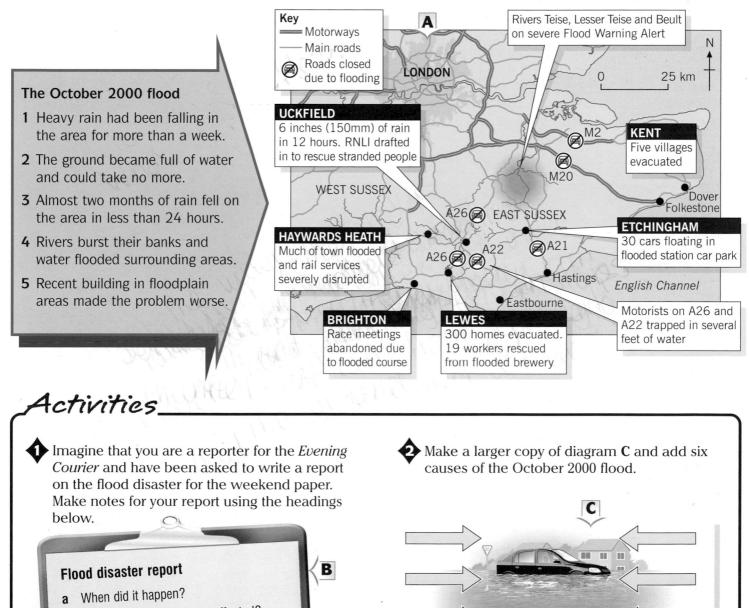

Key
- Motorways
- Main roads
- Roads closed due to flooding

A

LONDON

Rivers Teise, Lesser Teise and Beult on severe Flood Warning Alert

N

0   25 km

UCKFIELD
6 inches (150mm) of rain in 12 hours. RNLI drafted in to rescue stranded people

WEST SUSSEX

M2

M20

KENT
Five villages evacuated

Dover
Folkestone

A26   EAST SUSSEX

HAYWARDS HEATH
Much of town flooded and rail services severely disrupted

A26   A22   A21

ETCHINGHAM
30 cars floating in flooded station car park

Hastings

English Channel

Eastbourne

Motorists on A26 and A22 trapped in several feet of water

BRIGHTON
Race meetings abandoned due to flooded course

LEWES
300 homes evacuated. 19 workers rescued from flooded brewery

Activities

1 Imagine that you are a reporter for the *Evening Courier* and have been asked to write a report on the flood disaster for the weekend paper. Make notes for your report using the headings below.

Flood disaster report

a   When did it happen?

b   Which places were worst affected?

c   What were the main effects of the flood?

d   How were car owners and rail travellers affected?

e   What help was available to flood victims?

f   What problems will there be once the flood waters have gone down?

B

2 Make a larger copy of diagram **C** and add six causes of the October 2000 flood.

C

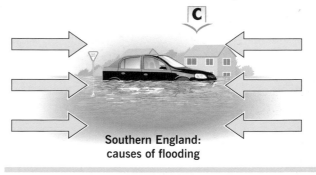

Southern England:
causes of flooding

Summary

Floods can cause much damage and seriously affect people's lives. There are usually several different causes of floods but some places are more at risk from flooding than others.

How does the UK cope with floods?

Flooding is a serious problem in the UK and it is happening more often. There are 2.3 million properties at risk and this number is expected to increase to over 5 million in the next 50 years. Autumn 2000 was the wettest since records began in 1766. Major flooding affected large parts of the country, and in some cases water levels were at their highest for over 100 years. Whilst flooding cannot be prevented, in rich countries like Britain much can be done to reduce the risk of floods and limit their worst effects.

The **Environment Agency** is an organisation that looks after rivers in England and Wales. It monitors rainfall, river levels and sea conditions 24 hours a day. This information is used to predict the possibility of flooding. If flooding is forecast, the Environment Agency's **Floodline** issues warnings. It also gives advice on what to do before, during and after a flood.

A   Planning for flooding in the UK

1 Study the UK's rivers and coasts and identify areas most at risk and where flooding would do most damage.

2 Recommend the building of flood defences such as embankments and overflow channels where they are needed.

3 Continually check rainfall and water levels to see if a river is going to flood.

4 Issue warnings through radio, TV and home visits for those in most danger, when floods are expected.

5 Alert emergency services such as the police, fire brigade and army, to provide help for those in need.

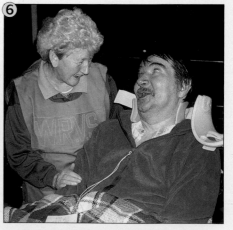

6 Ensure that food and shelter is available for those made homeless. Emergency medical care should also be available.

> " Flooding – you can't prevent it, but you can prepare for it. "

ENVIRONMENT AGENCY

Flood warning codes

Flood Watch means flooding is possible. Be aware! Be prepared! Watch out!

Flood Warning means flooding of homes, businesses and main roads is expected. Act now!

Severe Flood Warning means serious flooding is expected. There is imminent danger to life and property. Act now!

All Clear means there are no longer flood watches or flood warnings in force. Seek advice to return.

What to do in a flood

Before a flood

Be alert for flood warnings and take action.
Check on family and nearby neighbours.
Move people, pets and valuables to safety.
Collect warm clothes, food and a torch.
Block doorways with sandbags.
Switch-off electricity and gas.

During a flood

Listen to the local radio for flood news.
Never walk, drive or swim through flood water.
Avoid flood water as it may be contaminated.

After a flood

Check if it is safe to turn electricity and gas on.
Open windows and doors for ventilation.
Throw out contaminated food.
Wash taps and run them before use.
Clear up by disinfecting walls and floors.
Beware of rogue traders offering to help.
Call your insurance company for advice.

Activities

1 a Describe three ways that the Environment Agency can help reduce the risk of flooding.

b Describe three ways that the Agency can help limit the worst effects of flooding.

2 Which flood warning would have been given for the south of England floods of October 2000 (pages 48 and 49)? Give reasons for your answer.

3 'Floodline' encourages people to make a family flood plan like this one. Write out the plan and add a reason for each point.

Family Flood Plan
- Know how to contact each other.
- Put together an emergency flood kit.
- Know how to turn off power supplies.
- Put emergency numbers in a safe place.
- Understand the flood warning system.
- Listen to the local radio programme.

4 a Find out more about Floodline by telephoning for a Flood Warning Pack or using their website.

b Design a leaflet to give to people living in areas where there is a flood risk. Explain to them briefly what information is available and what they should do.

ENVIRONMENT AGENCY
Floodline
Telephone: 0845 988 1188
Website: www.environment-agency.gov.uk/flood

Summary

There is no easy way to cope with floods. Rich countries like the UK can afford schemes that help reduce the damaging effects of flooding.

Floods in Bangladesh, 2004

সুরমা **SURMA NEWSWEEKLY** **Dhaka 28 July 2004**

Floods hit Bangladesh yet again

The floods in Bangladesh are the worst since 1998. They have devastated two-thirds of the country, affecting 22 million people and making over 4 million homeless. The official death toll is 1,679 although many more are still missing. Most of the deaths were caused by drowning, lightning, poisonous snakes that slither through the water during flooding, and outbreaks of water-borne diseases.

Most of Dhaka, the country's capital city, is under water. Residents waded through the waist-deep flood waters holding their belongings over their heads. Small wooden boats and cycle rickshaws formed traffic jams whilst electrical wires dangled dangerously over some roads. The water turned blackish and foul-smelling as it mixed with raw sewage.

Reservoirs were polluted and gas outlets swamped, causing shortages of clean water and cooking fuel. The threat of disease is increasing and hospitals are already full of people suffering from dysentery and diarrhoea. Many children have developed fevers, coughs and rashes.

The countryside areas have also been badly hit. Many families are just recovering from the floods of six years ago. Once again they have lost their homes, lost their land and lost their cattle. Their crops have been ruined and they have no food or money. As in 1998, they desperately need help if they are to survive.

What caused the Bangladesh flood?

Bangladesh is a country in Asia. It is located at the mouth of two of the world's longest rivers, the Ganges and the Brahmaputra.

Bangladesh has floods every year – but they seem to be getting worse. The country relies on the heavy monsoon rains to flood the rice fields, but too much rain can destroy the crop as well as the homes of the farmers. In four monsoon months, Bangladesh can get as much rain as London gets in two years!

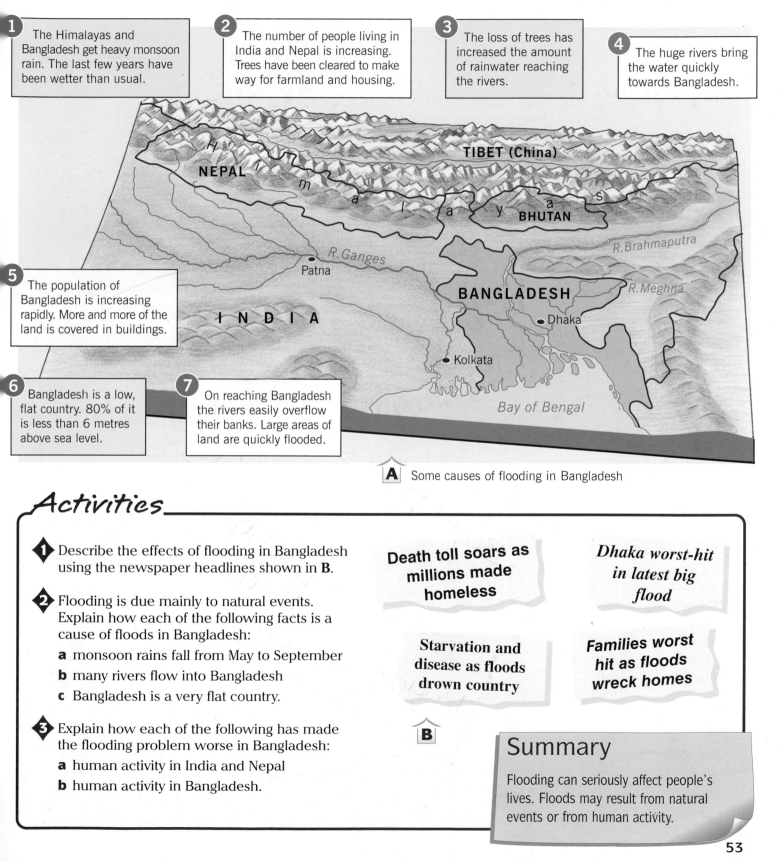

1 The Himalayas and Bangladesh get heavy monsoon rain. The last few years have been wetter than usual.

2 The number of people living in India and Nepal is increasing. Trees have been cleared to make way for farmland and housing.

3 The loss of trees has increased the amount of rainwater reaching the rivers.

4 The huge rivers bring the water quickly towards Bangladesh.

5 The population of Bangladesh is increasing rapidly. More and more of the land is covered in buildings.

6 Bangladesh is a low, flat country. 80% of it is less than 6 metres above sea level.

7 On reaching Bangladesh the rivers easily overflow their banks. Large areas of land are quickly flooded.

A Some causes of flooding in Bangladesh

Activities

1 Describe the effects of flooding in Bangladesh using the newspaper headlines shown in **B**.

2 Flooding is due mainly to natural events. Explain how each of the following facts is a cause of floods in Bangladesh:
  a monsoon rains fall from May to September
  b many rivers flow into Bangladesh
  c Bangladesh is a very flat country.

3 Explain how each of the following has made the flooding problem worse in Bangladesh:
  a human activity in India and Nepal
  b human activity in Bangladesh.

Death toll soars as millions made homeless

Dhaka worst-hit in latest big flood

Starvation and disease as floods drown country

Families worst hit as floods wreck homes

B

Summary

Flooding can seriously affect people's lives. Floods may result from natural events or from human activity.

How does Bangladesh cope with floods?

Bangladesh suffers more from flooding than any other country in the world. The problem is made worse because of the extreme poverty of the people who live there.

In 1989, after a particularly bad flood, several wealthy countries joined with Bangladesh to set up the Flood Action Plan. Under the Plan, billions of dollars are being spent on schemes which it is hoped will reduce the risk and danger of flooding. Some of the main points of the Plan are shown below.

A   Flood Action Plan for Bangladesh

Build **dams** to control river flow and hold back the monsoon rainwater in reservoirs. Stored water can be used for irrigation and to generate cheap electricity.

Build **embankments** and deepen river channels to stop the river overflowing. The embankments would be up to 7 metres high in urban areas.

Build 5,000 **flood shelters** in areas most at risk. These would be cheap to construct and provide a place of safety for almost everyone. They would be well stocked with food.

Improve **flood warning systems**. These would give early warnings of floods. They would also give instructions to people as to what they should do before, during and after the flood.

Provide **emergency help** when the floods arrive. Embankments would be repaired, people taken to safety and food and medical care provided to those in need.

Give **after-care** once the flood ends. Food, drinking water, tents, medicines and money would be available. Help would be given to plant seeds for next year's crops.

There is no easy solution to Bangladesh's flooding problem. The enormous size of the problem and the shortage of money make the task almost impossible.

Even the Flood Action Plan has not been welcomed by everyone. Many people are worried that such a large scheme could actually make the problem worse.

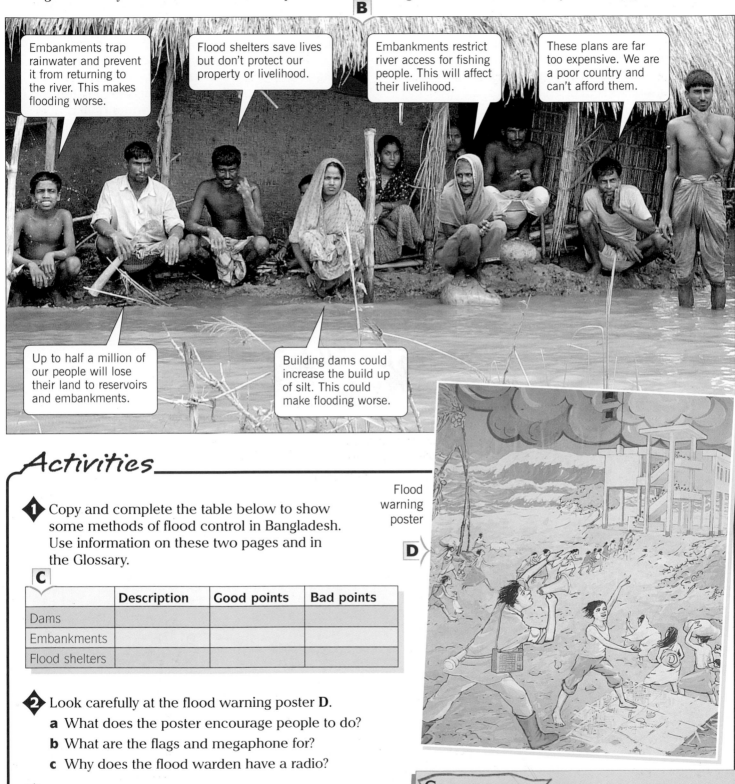

B

Embankments trap rainwater and prevent it from returning to the river. This makes flooding worse.

Flood shelters save lives but don't protect our property or livelihood.

Embankments restrict river access for fishing people. This will affect their livelihood.

These plans are far too expensive. We are a poor country and can't afford them.

Up to half a million of our people will lose their land to reservoirs and embankments.

Building dams could increase the build up of silt. This could make flooding worse.

Flood warning poster

D

Activities

1 Copy and complete the table below to show some methods of flood control in Bangladesh. Use information on these two pages and in the Glossary.

C

	Description	Good points	Bad points
Dams			
Embankments			
Flood shelters			

2 Look carefully at the flood warning poster **D**.

   a What does the poster encourage people to do?

   b What are the flags and megaphone for?

   c Why does the flood warden have a radio?

3 Some people say that Bangladesh will never be able to cope with its flooding problem. Do you agree with this? Give reasons for your answer.

Summary

Poor countries like Bangladesh find it very difficult to cope with floods. The effects of flooding are therefore a lot worse than they would be for a rich country.

How can the risk of flooding be reduced?

There are many different ways of controlling rivers and reducing the risk of flooding. The methods shown below are called **flood prevention schemes** because they try to stop floods happening.

Many people now believe that complete river and flood control is impossible. They say that flooding should be allowed to happen as a natural event. Flood prevention schemes can, in the long term, save money. They also improve water quality and help support wildlife.

A

Dams
A dam built across a river traps water and stores it in a reservoir. The water may then be released in a controlled way.

Forests
Trees may be planted in the drainage basin. These will slow down water movement and reduce the amount reaching the river.

Embankments
The river's banks may be built up with earth or concrete. This will make the river deeper and keep the water in.

Concrete linings
Line river channels in urban areas with concrete. This will take excess water quickly away from danger areas.

Activities

1 Draw a star diagram to show eight ways of reducing the risk of flooding. Write a short sentence to describe each one.

2 Look at the different approaches to flood prevention. Which approach do you think:

a costs most

b costs least

c may drown farmland and houses

d uses up most land

e protects the natural environment?

Give reasons for your answers.

3 One approach to flooding is simply to allow rivers to flood naturally. For each of the people below say if they would be **for** or **against** this method. Give reasons for your answer.

Local farmer

Flood protection manager

Bird watcher

Summary

A variety of methods can be used to reduce the risk of floods, but there is no way to stop flooding completely. A modern approach is to allow parts of a river to flood naturally.

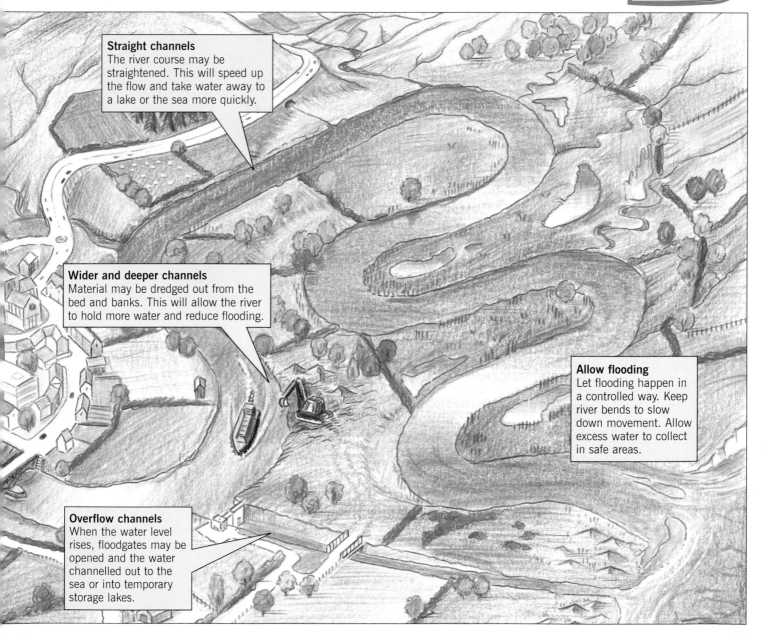

Straight channels
The river course may be straightened. This will speed up the flow and take water away to a lake or the sea more quickly.

Wider and deeper channels
Material may be dredged out from the bed and banks. This will allow the river to hold more water and reduce flooding.

Allow flooding
Let flooding happen in a controlled way. Keep river bends to slow down movement. Allow excess water to collect in safe areas.

Overflow channels
When the water level rises, floodgates may be opened and the water channelled out to the sea or into temporary storage lakes.

The river flooding enquiry

How should the Doveton valley be protected from flooding?

Look at map **D**, which shows part of the Doveton valley. In most years, the river overflows its banks and causes serious damage. The people in the area want their homes and land to be protected from the flooding. The Environment Agency has agreed to look at the problem. It has made a study of the area and suggested four different schemes to help stop the flooding. It is your task to decide which is the best scheme.

Factors to consider	Scheme A	Scheme B	Scheme C	Scheme D
Prevents all flooding				
Stops flooding in Crofton				
No homes lost				
No roads submerged				
No grazing land lost				
No good farmland lost				
Helps with irrigation				
Helps protect wildlife				
Not too expensive				
Total				

A

B

1
a Copy table **A** which shows some factors that have to be considered when choosing a flood protection scheme.

b Look carefully at the map and scheme descriptions. Show the advantages of each scheme by putting ticks in columns **A**, **B**, **C** or **D**. Complete one factor at a time. More than one column may be ticked for each factor.

c Add up the ticks to find which scheme has the most advantages.

d Which scheme would *you* choose? The one with the most advantages would be the best. If two schemes are equal, think about which parts of the valley you would want to protect most.

e Briefly describe the scheme you have chosen. Explain how it will help protect the valley from flooding.

2 The flood protection scheme will affect different people in different ways. Work in pairs and discuss what the people in the drawing below will think of your chosen scheme. For each person say if they would be **for** or **against** the scheme. Give reasons for their views.

C

Trudy Trout, owner of Crofton caravan park

Farmer Wally Wade of Hillside Farm

Barry Beer, owner of the Crofton Inn

Larry Laugh, local lorry driver

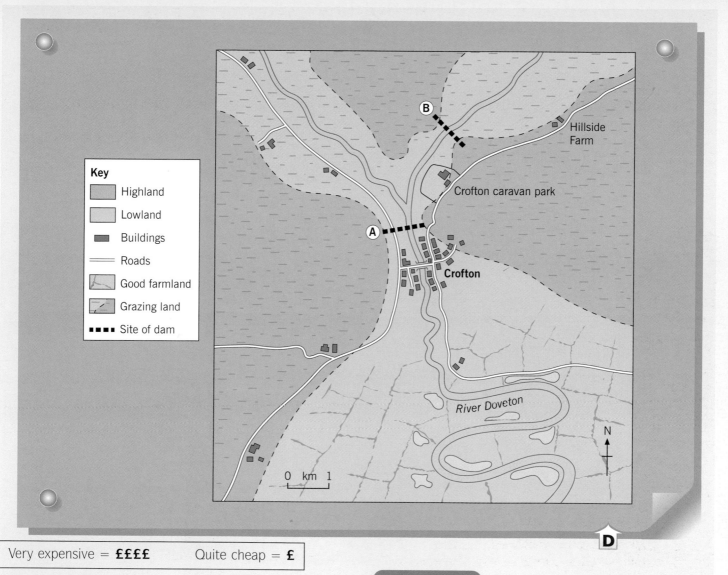

Very expensive = **££££**   Quite cheap = **£**

Scheme A

Build a dam at A and create a large reservoir above the village. Much farmland and several farms would be flooded. The scheme would stop flooding in the village and protect most of the valley.   **Cost = ££££**

Scheme B

Build a dam at B and create a small reservoir higher up the valley. Nobody would lose their home but some grazing land used by sheep and cattle would be lost. There would still be some flooding in Crofton and further downstream.   **Cost = ££**

Scheme C

Build a dam at B and deepen the river channel through Crofton. This would allow the water flowing through the village to move away more quickly. The scheme would protect Crofton but there may still be some flooding downstream.   **Cost = £££**

Scheme D

Build embankments at Crofton. Deepen and straighten the river below the village to take water away quickly. Allow natural flooding to happen downstream at the river bends. There would still be some flooding, especially upstream of Crofton.
   Cost = £

What are settlements like?

What is this unit about?

This unit is about the location, growth and nature of settlements. It looks at the functions and land use of settlements and shows how changes may bring benefits but can also cause problems.

In this unit you will learn about:

◆ how sites for settlements were chosen
◆ the benefits and problems of settlement growth
◆ land use patterns in towns
◆ how functions and land use change
◆ how shopping has changed
◆ traffic problems and solutions
◆ how environments may be improved.

A Central Birmingham

Why is learning about settlements important?

Most of us live in a settlement of some kind. What settlements are like, therefore, affects us all in some way or another. This unit can help us understand what settlements are about and how they try to provide for the needs of people living in them. It can also help us see how changes to settlements can directly affect us, and have an impact on the way we live our lives.

This unit can also help you in other ways. It can:

◆ give you an interest in the place where you live

◆ help you choose where you might want to live in the future

◆ help you appreciate the problems facing town planners

◆ help you choose the best transport to use

◆ help you understand how the environment may be improved in the area where you live.

B | Durham

> ◆ What might have attracted people to place B?
>
> ◆ For places A and C
>   – what do you think are their good points and bad points
>   – in which one would you prefer to live?

C | East End of London

How were the sites for early settlements chosen?

When we use the word **site** we mean the actual place where a village or town grew up. A site was chosen if it had one or more natural advantages.

Diagram **A** shows eight natural advantages. The more natural advantages a place had the more likely it was to grow in size.

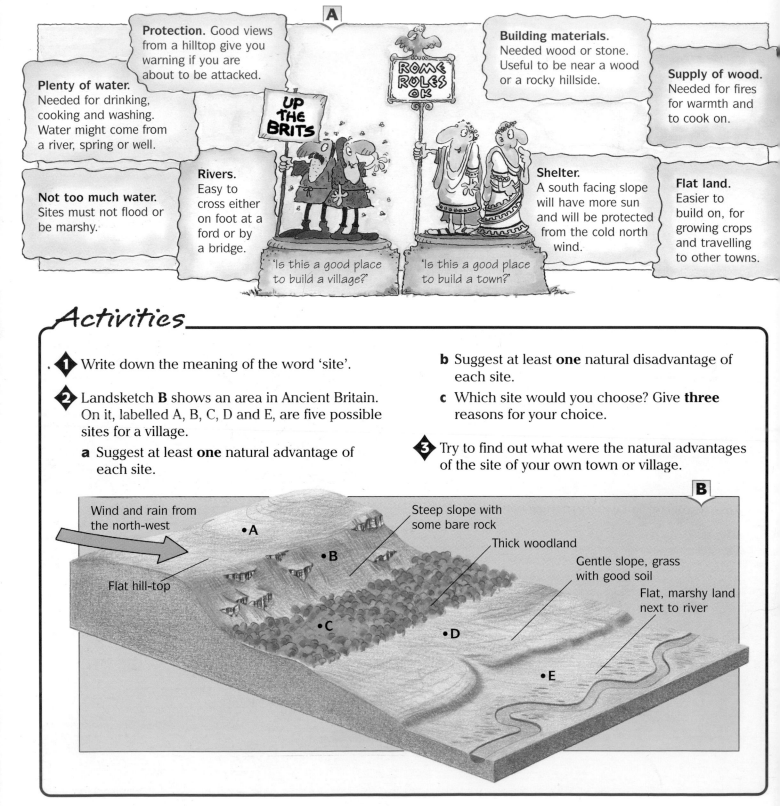

A

Protection. Good views from a hilltop give you warning if you are about to be attacked.

Plenty of water. Needed for drinking, cooking and washing. Water might come from a river, spring or well.

Rivers. Easy to cross either on foot at a ford or by a bridge.

Not too much water. Sites must not flood or be marshy.

Building materials. Needed wood or stone. Useful to be near a wood or a rocky hillside.

Supply of wood. Needed for fires for warmth and to cook on.

Shelter. A south facing slope will have more sun and will be protected from the cold north wind.

Flat land. Easier to build on, for growing crops and travelling to other towns.

'Is this a good place to build a village?'

'Is this a good place to build a town?'

ROME RULES OK

UP THE BRITS

Activities

1 Write down the meaning of the word 'site'.

2 Landsketch **B** shows an area in Ancient Britain. On it, labelled A, B, C, D and E, are five possible sites for a village.

  a Suggest at least **one** natural advantage of each site.

  b Suggest at least **one** natural disadvantage of each site.

  c Which site would you choose? Give **three** reasons for your choice.

3 Try to find out what were the natural advantages of the site of your own town or village.

B

Wind and rain from the north-west

Flat hill-top

• A

• B

• C

Steep slope with some bare rock

Thick woodland

• D

Gentle slope, grass with good soil

Flat, marshy land next to river

• E

The photo below shows Warkworth, a small village in Northumberland. It is located on a bend of the River Coquet. Early settlers were most concerned about their safety and getting a supply of food and water. As you can see, the site at Warkworth provided those needs.

Despite its good site, Warkworth has never grown into a large town. This is because the original advantages are no longer so important. Nowadays, people want to be near employment and services such as schools and hospitals. These are not so readily available at Warkworth.

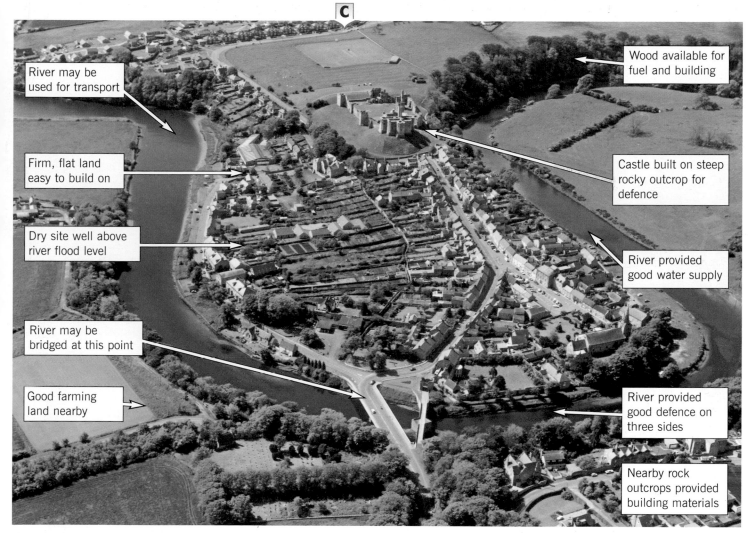

C

- River may be used for transport
- Firm, flat land easy to build on
- Dry site well above river flood level
- River may be bridged at this point
- Good farming land nearby
- Wood available for fuel and building
- Castle built on steep rocky outcrop for defence
- River provided good water supply
- River provided good defence on three sides
- Nearby rock outcrops provided building materials

Activities

1 Draw a star diagram like the one below to show the advantages of Warkworth as a site for a settlement. Give two advantages under each heading.

Defence — Food and water — Site advantages of Warkworth — Building materials — Building land

2 Complete table **D** to show how **some** of Warkworth's original site advantages are no longer so important.

D

Original advantage	Why no longer important
•	
•	

Summary

Early sites for settlements were chosen because of natural advantages such as a good water supply, dry land, defence, shelter, farmland and building materials.

What different settlement patterns are there?

If you look at map **D** on the next page you will see that the settlements have different shapes. Some are long and thin, some are compact and almost round, others are broken up and spread out. In geography we call these different shapes the **settlement pattern**.

Settlement patterns are usually influenced by the natural features of the area. These are often the same features that were considered important when choosing the original site for the settlement. The three main types of settlement pattern are shown below.

♦ A **dispersed settlement** has buildings that are well spread out.

♦ Settlements with this pattern are often found in highland areas where it is not easy to build houses close together. Here, people also needed more land to grow their crops or graze their animals.

Dispersed **A**

♦ A **nucleated settlement** has buildings closely grouped together.

♦ Settlements with this shape often grew around a road junction or river crossing. A long time ago people built their houses close together for safety. This pattern is common in lower, flatter parts of Britain.

Nucleated **B**

♦ **Linear settlements** are often called **ribbon developments** because they have a long, narrow shape.

♦ Settlements with this shape usually grow along a narrow valley where there is little space. They may also be found strung along a road or on either side of a river.

Linear **C**

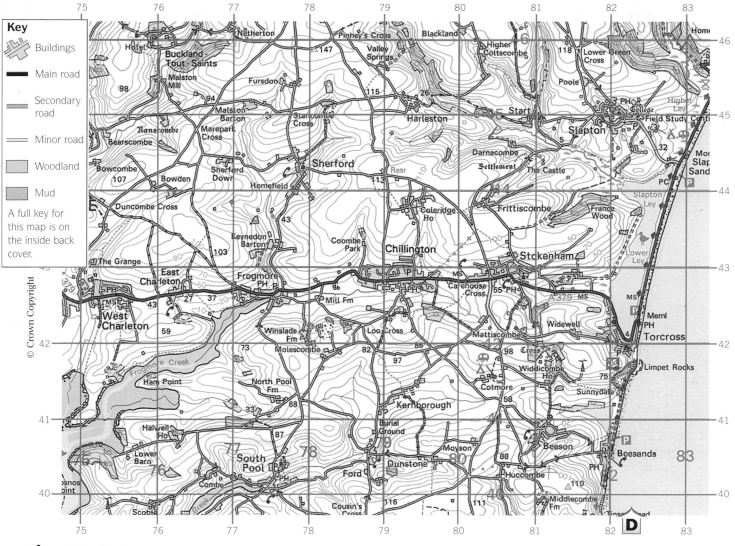

Activities

1 Copy the settlement pattern drawings below. Label each one **dispersed**, **nucleated** or **linear**. Write a brief description of each one. Suggest a reason for its shape.

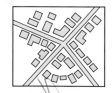 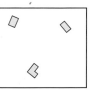

2 Map **D** is part of Devon in south-west England. It shows many different settlement patterns.

  a Make a larger copy of table **F**.

  b Complete your table by filling in the empty boxes. The first one has been done to help you. (You may need to look at page 132 to remind you about grid references.)

  c Find another example of a nucleated settlement and a linear settlement and add them to your table.

Village name	Map reference	Simple drawing	Settlement pattern
Bowden	7644		Dispersed
Slapton	8144		
South Pool	7740		
Cotmore	8041		
Beeson			
Torcross			
Sherford			

Summary

The three main types of settlement pattern are dispersed, nucleated and linear. The shape of a settlement is usually determined by the physical features of the surrounding area.

How do settlements change with time?

No town or village remains the same for ever. Over a period of time the following may all change:

1 the **shape** of a settlement
2 the **function** of a settlement
3 the **land use** of a settlement
4 the **number** and **type** of people living in the settlement.

Villages are small in size so it is often easier to see these changes in them than it is to see changes in a large town or city.

What was a typical village like in the 1890s? Although no two villages are the same, most have several things in common. Diagram **A** shows a typical village about one hundred years ago. In the centre there was often a village green. Buildings were grouped closely together (nucleated) around this green forming a core. Roads were usually narrow lanes. Most houses were small terraced cottages. The people who lived in them would probably have been born in the village. Most would have worked on local farms. As houses and farms were built at different times they would have different styles and building materials.

How had the village changed by the 2000s? Diagram **B** shows the same village today. The village has grown larger and has many new buildings. It has become **suburbanised**. This means it has become similar to the outskirts of larger towns.

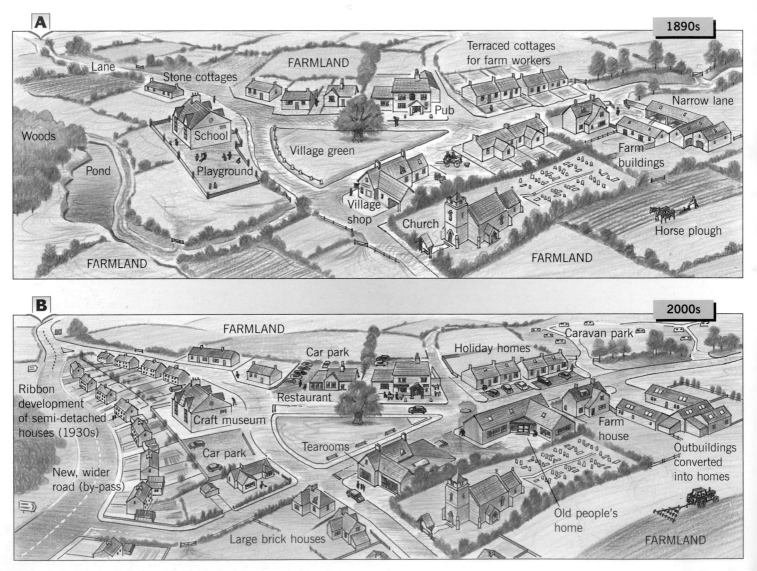

Activities

1 Write down the meaning of:

a shape

b function

c land use

when talking about a settlement.

2 **Spot the differences!** List at least **ten** differences between the village in the 1890s and the village in the 2000s.

3 The changes to the village will have affected different groups of people in different ways. Look at the pictures of some of these groups of people shown below. Match up the pictures with the statements below numbered **1** to **8**.

For example:
Young married couple = statement **2**

4 Activity **2** asked you to find the differences in the village between the 1890s and the 2000s. Why do you think changes have been made in:

a the number and type of houses

b the use of buildings around the green

c the use of the land around the village

d the roads?

5 It has been suggested that the woods should be cleared so that an estate of expensive houses can be built.

a Which groups of people will like this change?

b Which groups of people will be against this change?

Suggest reasons for your answers.

How groups may be affected

1 I might have to close as most people have cars to shop in town.

2 We are just married and cannot afford an expensive house.

3 The extra noise frightens away the wildlife.

4 To get customers I have to provide food for townspeople. Villagers only want a drink.

5 I made money by selling my land so that houses could be built. Now people walk on the land I still own.

6 With all the new houses I have plenty of work to do.

7 I have to travel 10 km to school. At night there is nothing to do.

8 I came here for peace and quiet. Now I cannot drive into town and there are no buses.

Farmer · Shopkeeper · Bird watcher · Teenager · Young married couple · Restaurant owner · Elderly person · Builder

Summary

Settlements change over a period of time.
These changes can affect:
- the size and shape of the settlement
- the environment, e.g. new roads, larger villages
- the lives of people living in the settlement.

What are the benefits and problems of settlement growth?

In Britain most people are urban dwellers living in towns and cities. These settlements grew very quickly in the nineteenth century. This was when large numbers of people moved there to work. Today, Britain's cities are no longer growing in size. However, in many overseas countries people are still moving to cities in large numbers.

This is because they believe that many **benefits** come from living and working in cities. Moving there will improve their **quality of life**.

Drawing **A** shows some of the benefits which people hope to find in large towns and cities.

Benefits

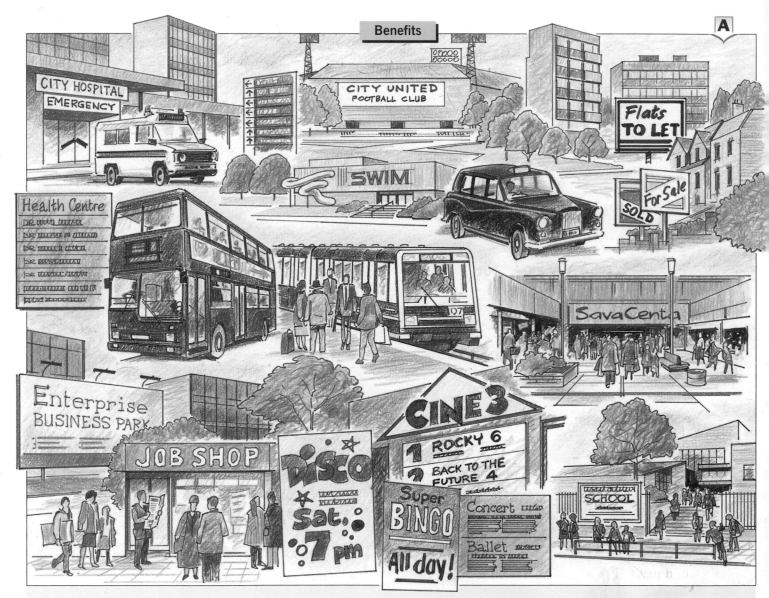

- There are more houses and flats to buy or to rent.
- There are more jobs which are better paid.
- Food supplies are more reliable, with many shops giving a greater choice.

- It takes less time and money to travel to work and to shops.
- There are more and better services, such as schools and hospitals.
- Urban areas have 'bright light' attractions, such as discos, concerts and sporting activities.

For people already living in cities, life is often less attractive. Living and working in cities creates many **problems**. Drawing **B** shows some of the problems found in British cities.

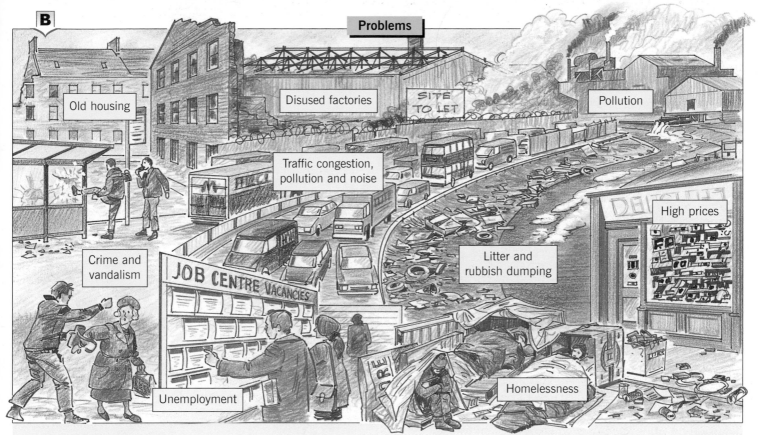

B

Problems

Old housing

Disused factories

SITE TO LET

Pollution

Traffic congestion, pollution and noise

High prices

Crime and vandalism

JOB CENTRE VACANCIES

Litter and rubbish dumping

Unemployment

Homelessness

◆ Traffic causes congestion, accidents, noise and air pollution.

◆ Old roads are too narrow for lorries and buses; new roads take up much land.

◆ Old houses and factories need urgent, expensive repairs or they are left empty.

◆ There is waste land where houses and factories have been pulled down.

◆ Crime, vandalism and litter make cities dangerous and unpleasant.

◆ Land is very expensive to buy, in and near the city centre.

Activities

1 a Make a copy of the table below. List the **three** things which you think are best about living in cities, and the **three** things you think are the worst.

Cities	
Good news	Bad news

b Do you think there is more good news or bad news?

2 If you had to move from where you live, would it be to a bigger or a smaller settlement? Give reasons for your answer.

3 Try to find out what has been done in your local town or city to try to reduce:

a traffic problems

b pollution

c crime, vandalism and litter.

4 Suggest other ways in which these three problems may be overcome.

Summary

Many people move to large cities because they see benefits in living and working there. However, as these settlements become older and bigger, many problems are created.

Why are there different land use patterns in towns?

When each town first began to grow it usually had one particular use or function. Towns and cities of today often have several different functions. The main functions are **commerce** (shops and offices), **industry** (factories), **residential** (flats and houses) and **open space** (parks and sports facilities). As each function tends to be found in a particular part of a town, then a pattern of land use develops. Although no two towns will have exactly the same pattern of land use, most have similar patterns. When a simple map is drawn to show these similarities it is called an **urban model**.

The model in diagram **A** shows four differences in land use drawn as a series of circles around the city centre. It is suggested that this pattern developed for two reasons:

1 The oldest part of a town is in the middle. As the town grew, larger new buildings were built on the edges.

2 Land in the city centre is expensive to buy. This is because many different types of land users would like this site and so they compete for it. Usually the price of land falls towards the edges of towns.

Activities

1 a Copy diagram **A** and colour in the four zones.

 b Name the four zones.

 c Name an area in your local town for each of the zones.

2 Diagram **B** shows four houses that are for sale. In terms of location, cost and amenities, which house might you choose:

 a if you were a first-time home buyer

 b if you had two children aged under 6 years

 c when your children leave home and you have a good job

 d when it is time for you to retire from work

 e if you could choose for yourself as a teenager?

 In each case give reasons for your answer.

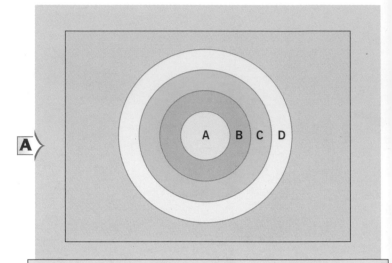
A

Zone A The central business district (CBD)
The centre of the town was the first place to be built. It is still full of shops, offices, banks and restaurants. There are very few houses and little open space here.

Zone B The inner city
This used to be full of large factories and rows of terraced housing built in the nineteenth century. Houses were small and there was no open space as land was expensive. Today most of the big factories have closed and the oldest houses have been replaced or modernised.

Zone C The inner suburbs
This is mainly semi-detached housing built in the 1920s and 1930s. There is some open space.

Zone D The outer suburbs
This includes large, modern houses and some council estates built since the 1970s. Recently small industrial estates, business parks and large supermarkets have been built here. There are large areas of open space.

Summary

The main function, or land use, of an area may result from its age and the cost of land.

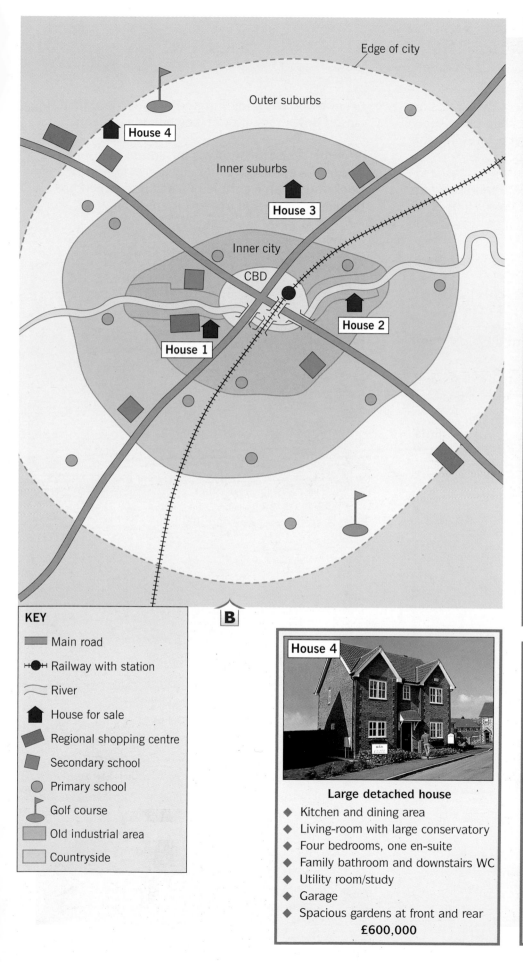

KEY

- Main road
- Railway with station
- River
- House for sale
- Regional shopping centre
- Secondary school
- Primary school
- Golf course
- Old industrial area
- Countryside

Edge of city

Outer suburbs

House 4

Inner suburbs

House 3

Inner city

CBD

House 2

House 1

B

House 1

Mid-terraced

- Kitchen
- Living room
- Two bedrooms
- Bathroom extension in backyard
- In need of some modernisation

£100,000

House 2

Modern fifth-floor apartment/flat in a recently converted factory

- Modern kitchen
- Living room with dining area
- Two bedrooms
- Bathroom
- Parking space

£180,000

House 4

Large detached house

- Kitchen and dining area
- Living-room with large conservatory
- Four bedrooms, one en-suite
- Family bathroom and downstairs WC
- Utility room/study
- Garage
- Spacious gardens at front and rear

£600,000

House 3

Semi-detached in a cul-de-sac

- Kitchen
- Living room and dining room
- Three bedrooms
- Bathroom and separate WC
- Garage and small conservatory
- Garden front and rear

£300,000

Why does land use in towns change?

Land use in towns changes over time. City centres are modernised to attract more people, while open space on the outskirts is turned into housing estates, large shopping centres and industrial parks.

Most **inner city** areas were built over a hundred years ago. Naturally they have aged in that time. Houses became too old and cramped to live in and factories closed down. The inner cities had to change.

Over the years, most towns have tried to improve conditions in the old inner city areas. London Docklands is an example of a scheme which has brought change and improvement to an area.

A

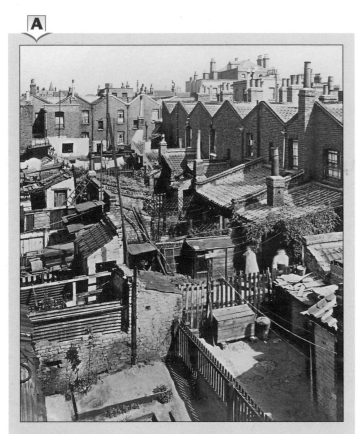

B

London Docklands – the problem

◆ Up to the early 1950s, London was the busiest port in the world and the Docklands a thriving industrial zone.

◆ For a number of reasons, shipping on the Thames went into decline, and by 1981 the docks were virtually abandoned and derelict.

◆ By that time there were very few jobs, transport was poor and there was a lack of basic services.

◆ Housing was a particular problem, with many old terraced houses lacking a bathroom or indoor toilet and in need of urgent repair.

◆ Although conditions were difficult, a strong 'EastEnders' community spirit built up.

London Docklands – the solution

◆ In 1981 the London Dockland Development Corporation was set up with the aim of improving living and working conditions.

◆ It began by clearing the old docks and houses and turning warehouses into expensive flats.

◆ Old industries were replaced with those using high-technology, such as newspapers, and by office blocks of financial firms.

◆ Underground stations were improved and a new City Airport and Docklands Light Railway built.

◆ The environment has been improved, trees planted and new parklands created.

How has the development affected people?

Many people have benefited from the Docklands redevelopment and are in favour of the scheme. Others, however, are less happy and are against it. Local people particularly feel disadvantaged. They say that housing is too expensive for them, that money

has been spent on facilities for the rich rather than the poor, and that most jobs are inappropriate to their needs. They also think that the 'yuppie' newcomers rarely mix with local people and that the 'EastEnders' community spirit has been broken up.

C London Docklands – some winners and losers

Young married couple We will have to move as we cannot afford to buy a place to live. A cheap flat is over £180,000.

Local shopkeepers All these newcomers mean more trade – especially as they have plenty of money to spend.

Financial manager We have modern offices and there is good-quality housing here. It only takes 10 minutes to travel into central London.

Local people Most new jobs go to highly skilled people from outside the area. Our close-knit community has been broken up.

Elderly people Shopping is expensive. Money is spent on houses and offices, not on hospitals and old people.

School leavers We are being trained to use computers. We will be able to get jobs and stay in the area.

Activities

1 a What was the early success of Docklands based on?

b What was the reason for its decline?

c What problems were caused by this decline?

d Describe a typical house shown in photo **A**.

e What is meant by 'a good community spirit'?

2 Make a copy of diagram **D** below and use the headings to describe the changes to London Docklands.

Buildings — Industry/Jobs — Transport — Environment **D**

3 Look carefully at drawing **C** above.

a Which people do you think are winners?

b Which people do you think are losers?

Give reasons for your answers.

4 Overall, do you think the changes to London Docklands have been good or bad? Give reasons for your answer – but remember to think about different points of view.

Summary

As time passes, the functions and land uses of different parts of a town will change. These changes affect different groups of people in different ways.

73

Where do we shop?

We all go shopping. We need to buy things so that we can feed ourselves and live our lives. Some of us shop simply for enjoyment. Recent surveys have found that shopping is one of Britain's most popular leisure activities.

Shopping is also big business. Each year, people in the UK spend over £38 billion on food shopping alone. A further £33 billion is spent on clothes and shoes. More than 2 million people in the UK work in shops, many of them part-time.

As we have seen, smaller settlements usually have very few shops. Larger settlements, however, are likely to have several shopping centres. In recent years, many of these have been built out of town and well away from the city centre.

Where people choose to shop depends upon what they want to buy and how often they need that product. The larger the shopping centre the more choice of goods and services there will be. People will travel long distances to these centres.

♦ Some **goods** such as food and newspapers don't cost very much and may be needed every day. We are happy to buy them in the nearest convenient place.

♦ These are called **convenience** or **low order goods**. They may be bought at the local **corner shop**, nearby shopping centre or supermarket.

A

♦ Some goods like clothes and furniture are much more expensive. We buy them less often and like to compare styles and prices before we buy.

♦ These are called **comparison** or **high order goods**. They are bought at large shopping centres, either in the city centre or out of town. Here, there is usually a good choice and lower prices because of competition.

B

Activities

1 Complete these sentences.
  a Convenience goods are ...
  b Comparison goods are ...
  c A corner shop is ...

2 Look at the goods shown in drawing **C**. Sort them into two groups: convenience goods and comparison goods.

3 Now decide where you would buy the goods in drawing **C**. Look at drawing **D**, which gives you three choices. You must make a separate trip to each centre. Write out a shopping list for each visit.

C

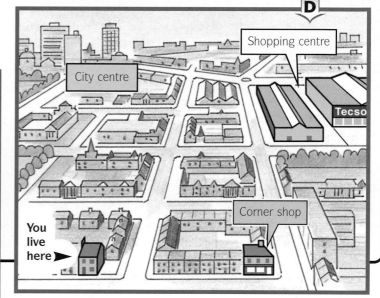

D

Shopping centre

City centre

Tecso

Corner shop

You live here ▶

EXTRA

You may be close enough to a shopping centre to visit it during a geography lesson. If you do you **must** go in groups and take great care when crossing roads.

Aim – to compare the shopping habits of people at your local shopping centre with those of people in the city centre.

Equipment – questionnaire, clipboard and pencil.

Method

a Make up a questionnaire similar to the completed one on the right.

b Politely ask at least 20 male and female shoppers of different ages the questions.

c Share your answers with other groups.

d Think of ways to illustrate your results.

e Describe your findings.

f Suggest reasons for differences between the results of your questionnaire and those of the one taken in the city centre.

Shopping survey

The numbers on the right show the result of asking 100 people in a city centre shopping area (mall) the following questions:

1 Do you shop here
  • every day? — 15
  • two or three times a week? — 15
  • once a week? — 50
  • once a month? — 20

2 Have you travelled
  • less than 1 mile? — 15
  • between 1 and 2 miles? — 20
  • between 2 and 5 miles? — 30
  • over 5 miles? — 35

3 Do you travel here
  • on foot? — 5
  • by car? — 75
  • by bus? — 15
  • any other way? — 5

4 Do you do most of your weekly shopping here?
  • Yes — 75
  • No — 25

5 What is the main thing you buy here?
  • Food — 30
  • Clothes — 40
  • Furniture — 10
  • Domestic equipment — 10
  • Others — 10

Summary

There are many different types of shopping centre. The larger the centre, the greater the choice of shops and goods there are to buy.

How has shopping changed?

The city centre is the main shopping area in a town. It has the largest number of shops, the biggest shops and the most shoppers. People are willing to travel long distances to the city centre because of the great choice of goods that they can buy there.

The main advantage of the city centre is its **accessibility**. Most of the main roads, bus routes and rail systems from the suburbs and surrounding areas meet at the city centre. It is therefore the easiest place for most people living in the town to reach.

City centre shopping has changed a lot recently. Attempts have been made to reduce traffic congestion, and to provide for the safety and comfort of shoppers. Many towns now have covered **shopping malls** which give protection from the weather and are traffic free.

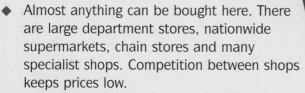

City centre

A

◆ Almost anything can be bought here. There are large department stores, nationwide supermarkets, chain stores and many specialist shops. Competition between shops keeps prices low.

◆ Good accessibility by car and public transport make it the most visited and busiest type of shopping centre.

◆ Overcrowding and traffic congestion cause problems. Pedestrianised areas and covered walkways have improved shopper comfort.

Activities

1 a Name the main street, shopping area or mall in your local town or city.

   b Name one department store, one nationwide supermarket and three specialist shops found in it.

2 Make a larger copy of sketch **B** showing a city centre shopping centre. Complete it by answering the questions.

B

f What is being done to improve city centre shopping?

e What are the main problems of these centres?

a What makes the city centre the main shopping centre?

b Why are people willing to travel long distances to this centre?

c Name four different types of shops in the city centre.

d What three things make the city centre accessible?

The biggest change in shopping has been the development of huge out-of-town shopping centres. These are located on the edge of cities, usually next to a main road or motorway. They are designed to attract motorists from a wide area and offer good accessibility and free parking.

Many people are worried about the development of these centres. They are concerned that they take trade away from the traditional shopping outlets. Some city centres have lost up to half of their business in the last ten years and are in serious decline. Many smaller shopping parades and corner shops have had to close altogether.

Another drawback of out-of-town shopping is the increased use of cars. This has caused more air pollution, noise and traffic congestion in suburban areas.

The government is very concerned about these effects. In future, it might not give permission to build any more out-of-town centres.

Out of town

C

- ◆ Ideal for shopping by car, with good road access and free car parking.

- ◆ Contain a large number of shops with a wide choice of goods. Prices are kept low by bulk buying and low running costs.

- ◆ Popular with shoppers who enjoy the bright and attractive air-conditioned shopping malls. Security staff ensure safety for families.

- ◆ Most centres have cafés, restaurants, cinemas and a wide range of other leisure amenities.

3 Design a poster to show the advantages of using a shopping centre like the one shown in photo **C**. Your poster should be attractive and interesting, and show the facts.

D

4 Imagine that you own a small shop close to a new out-of-town shopping centre. Your profits are down and you think you may soon have to close. Write a letter to the local council to say that the centre was a mistake. Mention all the ways that you think it is harming the area.

Summary

Shopping habits are changing. The city centre has always been the main shopping area in a town but it is now often congested and expensive. As more people shop by car, modern out-of-town centres are becoming increasingly popular.

How does internet shopping affect us?

We do not always have to go to shops to buy goods. We can stay at home and buy them over the telephone or on the **internet**. The internet is a network of computers around the world all linked together. As you know, you can connect your computer to it using the phone line. Once connected, you have an instant link with any place you want.

Using the internet you can now buy goods from places all over the world. Almost anything can be ordered **on-line**, from your weekly groceries to clothes, books, furniture and cars. They can then be delivered to your door, usually in a matter of days.

Internet shopping is now big business. As graph **A** shows, its value has increased rapidly in the last few years and this growth is expected to continue well into the future. The two main reasons for the popularity of internet shopping are convenience and low prices.

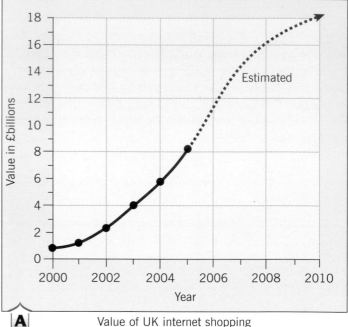

A Value of UK internet shopping

How internet shopping works

B

1 Connect to a company's website and choose the goods or services that you want.

2 Place your order and give your credit card number and address details for payment.

3 The order is sent instantly through the internet to the company.

4 The order is received by the company which then confirm details and costs.

5 The order is sent to the company's distribution warehouse in minutes.

6 The order is processed and packed for posting.

7 The order is transported by plane, van or lorry.

8 It is finally delivered to your door.

In 2005, 14.1 million households could access the internet from home. Of these, just over half had used on-line shopping to order goods or services. The most common purchases were for travel, accommodation and holidays. DVDs and music CDs were the next most popular purchase.

The change in shopping habits that the internet has brought about will have a considerable effect not only on the shopping industry, but also on the environment and the way we live our lives. Diagram **C** shows some of the advantages and disadvantages of internet shopping.

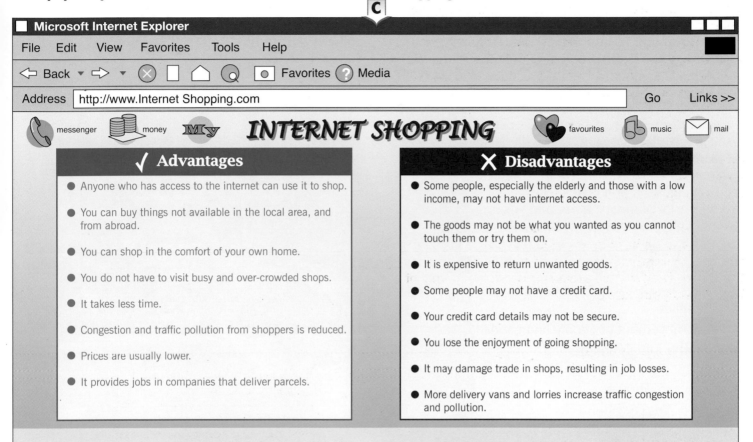

C

Microsoft Internet Explorer

File   Edit   View   Favorites   Tools   Help

Back   →   ⊗ ▢ ⌂ ◑ ● Favorites ? Media

Address | http://www.Internet Shopping.com | Go   Links >>

messenger   money   My   *INTERNET SHOPPING*   favourites   music   mail

✓ Advantages

- Anyone who has access to the internet can use it to shop.
- You can buy things not available in the local area, and from abroad.
- You can shop in the comfort of your own home.
- You do not have to visit busy and over-crowded shops.
- It takes less time.
- Congestion and traffic pollution from shoppers is reduced.
- Prices are usually lower.
- It provides jobs in companies that deliver parcels.

✗ Disadvantages

- Some people, especially the elderly and those with a low income, may not have internet access.
- The goods may not be what you wanted as you cannot touch them or try them on.
- It is expensive to return unwanted goods.
- Some people may not have a credit card.
- Your credit card details may not be secure.
- You lose the enjoyment of going shopping.
- It may damage trade in shops, resulting in job losses.
- More delivery vans and lorries increase traffic congestion and pollution.

Activities

1 Briefly describe how internet shopping works. Include the following in your answer:
- Transport   • Warehouse   • House delivery
- Company office   • Home computer

A simple diagram could help.

2 Look at each of the items in **D** below. Would you be happy to buy it on the internet? Give reasons for your answers.

D

- Tickets to a show
- Weekly food shopping
- Take-away meal
- Flowers
- Trainers
- Music CDs

3 How do you think internet shopping will affect the following? Give reasons for your answers.
- **a** A travel agent in a city centre
- **b** City centre air pollution
- **c** The number of jobs in shops
- **d** Motorway traffic
- **e** Someone living in a rural area
- **f** A young mother who works long hours

Summary

The internet has made shopping easier for many people. Its growth may affect the trade of traditional shopping outlets but it can also help reduce congestion and pollution in towns.

Traffic in urban areas – why is it a problem?

Traffic is a serious problem in most urban areas. Large cities such as London, Paris and New York have over a million cars trying to move around in their central areas. Movement is often impossible. Perhaps worse than congestion is the problem of pollution. Exhaust fumes are poisonous and can seriously damage health. Some city workers are so concerned that they wear masks to protect themselves. So what can be done? What are the causes of the problem, and why have they not been solved?

What is the problem?

Look at any urban area and you will soon be able to answer this question. Cars, buses and lorries all over the place cause congestion and chaos. They produce a lot of fumes, noise and danger. Other effects are:

- traffic jams blocking roads and stopping all movement
- delays for police, fire service and ambulances
- slow movement of people and goods
- loss of business and money
- people and buildings affected by noise and vibrations
- danger from accidents
- harmful exhaust fumes
- lack of parking places.

We live in an age of rapid transport yet vehicle movement is now actually slower than it was 80 years ago.

A

Activities

1 Look at the information with photo **A**. List what you think are the five worst problems caused by increased traffic in towns.

2 The two people in **B** are badly affected by traffic congestion and pollution. For each person write a letter to the local MP explaining how the problem affects them personally.

B

A businessman who lives out of town and drives to the city centre each day to work

A local resident with two young children who lives close to the main road

What is the cause?

There are many reasons for the traffic problems in our cities. The main one is simply that the number of cars has increased at a tremendous rate and there are now too many cars for cities to handle. It is predicted that this increase in cars will continue. By 2025 the number of cars might double and the number of lorries be three times greater than in 2000.

Another reason for traffic problems in cities is that most city centres were designed and built before cars were invented. They are therefore just not suited to today's transport. The problem is worst in the morning and in the late afternoon when people are travelling to and from work. This is called **the rush hour**.

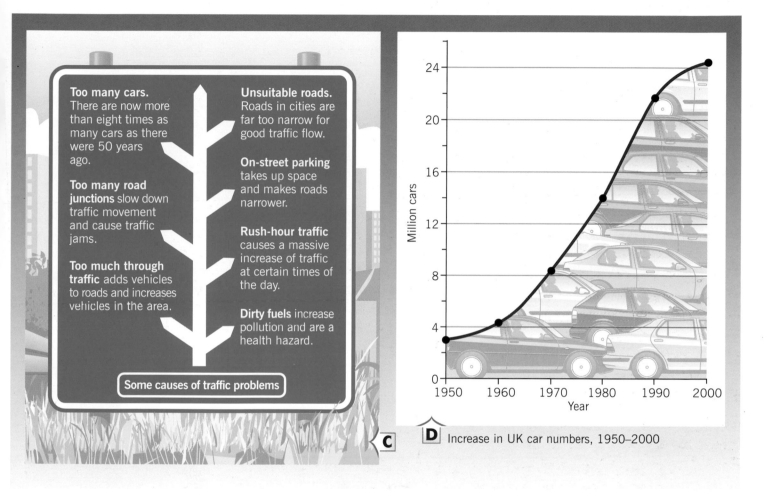

Too many cars. There are now more than eight times as many cars as there were 50 years ago.

Too many road junctions slow down traffic movement and cause traffic jams.

Too much through traffic adds vehicles to roads and increases vehicles in the area.

Unsuitable roads. Roads in cities are far too narrow for good traffic flow.

On-street parking takes up space and makes roads narrower.

Rush-hour traffic causes a massive increase of traffic at certain times of the day.

Dirty fuels increase pollution and are a health hazard.

Some causes of traffic problems

C

D Increase in UK car numbers, 1950–2000

3 Look at drawing **C**. List what you think are the five worst problems caused by increased traffic in cities. You need only write out the words in **bold**.

4 Look at graph **D**.

a How many cars were on Britain's roads in 1950? How many were there in 2000?

b Which of the graphs in **E** looks most like graph **D**? Use that description to describe the change in car numbers between 1950 and 2000.

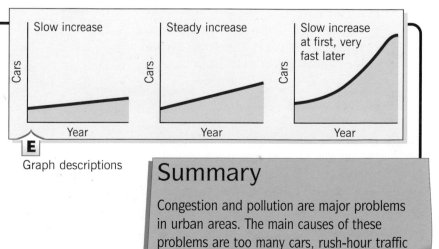

Slow increase

Steady increase

Slow increase at first, very fast later

E Graph descriptions

Summary

Congestion and pollution are major problems in urban areas. The main causes of these problems are too many cars, rush-hour traffic and unsuitable roads.

Traffic in urban areas – is there a solution?

There are two main ways of approaching the problem. The first is to allow **private transport** to increase and make improvements to cope with larger amounts of traffic. The second is to restrict private transport and discourage motorists from bringing cars into town centres. This would mean improving **public transport** such as bus and train services.

In fact, the traffic problem is so big and complicated that no single solution will ever completely solve it. The best way is to try to reduce the worst parts of the problem by using several solutions together. Some ideas that have been tried are shown in diagram **A**. Can you think of any others?

A

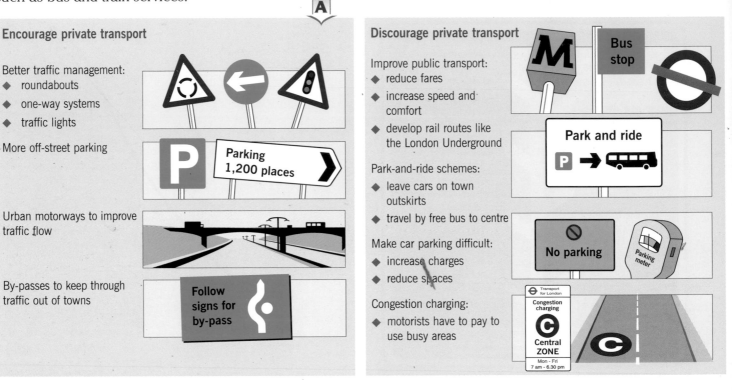

Encourage private transport

Better traffic management:
◆ roundabouts
◆ one-way systems
◆ traffic lights

More off-street parking

Parking 1,200 places

Urban motorways to improve traffic flow

By-passes to keep through traffic out of towns

Follow signs for by-pass

Discourage private transport

Improve public transport:
◆ reduce fares
◆ increase speed and comfort
◆ develop rail routes like the London Underground

Park-and-ride schemes:
◆ leave cars on town outskirts
◆ travel by free bus to centre

Make car parking difficult:
◆ increase charges
◆ reduce spaces

Congestion charging:
◆ motorists have to pay to use busy areas

Bus stop

Park and ride

No parking

Parking meter

Transport for London
Congestion charging
C
Central ZONE
Mon - Fri
7 am - 6.30 pm

Activities

1 **a** What is meant by public transport?

b What is meant by private transport?

c Name each of the following types of transport and sort them into **Public** and **Private**.

2 Draw a poster to discourage motorists from taking their cars into town centres.
• Show the bad things about town centres.
• Show the other types of transport that can be used.
• Colour your poster and make it interesting and attractive.

3 Diagram **B** shows how private and public transport in towns affect each other.

Copy and complete the diagram using the following phrases.

Fewer public transport users

Poorer-quality public transport

Increase in car use

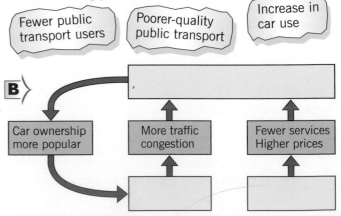

B

Car ownership more popular

More traffic congestion

Fewer services Higher prices

Public transport systems

In an attempt to reduce congestion and pollution, some cities have built new public transport systems. The aim of these systems is to move people around as safely, quickly and cheaply as possible. They are also designed to reduce pollution and so protect the environment.

Manchester's Metrolink tram system is an example of a surface light railway. This means that it runs on rails along existing roads and sections of former railway. Priority is given to the tram at traffic lights and road crossings. Travel by Metrolink has proven faster and cheaper to use than car transport.

The Manchester Metrolink

C

METROLINK
OPERATED BY SERCO METROLINK

- Connects suburbs to city centre
- Provides convenient way to travel in city centre
- Links with existing bus and rail routes
- Trams every 6 minutes during daytime
- Each tram can carry 206 passengers
- Fares subsidised to provide cheap travel
- Automatic ticket machines reduce costs

- Speeds up to 80 km/hr on former railways
- 26 stations and over 29 km of track
- Excellent wheelchair and pram access
- Powered by electricity so reduces noise and air pollution
- Carries over 16 million passengers a year
- Takes up to 2.5 million car journeys a year off the road

---

4 How does the Metrolink:

a provide a fast and cheap service

b serve people living away from the system

c help protect the environment?

5 Public transport is not popular with everyone. Make a list of the disadvantages of a system such as Metrolink.

6 Describe a traffic problem near your school or where you live. Suggest how the problem could be reduced.

Summary

Solving the problem of urban traffic is difficult. Better public transport may be the best way to improve people's movements without further damaging the environment.

Where should the by-pass go?

When a place becomes too crowded with vehicles (congested), a road can be built around it to take away some of the traffic. A road that is built to avoid a congested area is called a **by-pass**. Some by-passes are very long. The M25 which goes all the way round London is over 160 km (100 miles) long. Most by-passes are much shorter than this.

Building a by-pass is not easy. Money has to be found, suitable routes planned out, and discussions held between people whom the route may affect. This is all very difficult and takes a long time.

Activities

Look carefully at drawing **C** opposite. It shows the area around Haydon Bridge, a small town on the banks of the River Tyne between Newcastle and Carlisle.

The amount of traffic passing through Haydon Bridge has increased considerably in recent years. This has brought congestion and pollution as well as many accidents. Traffic hold-ups are also common on the busy A69 trunk road. It has been decided to build a by-pass to reduce the traffic going through the town centre.

Three possible routes have been suggested for the by-pass. Your task is to choose the best one.

1 **a** Copy table **A**, which shows some things that should be considered when choosing a by-pass route.

b Show the advantages of each route by putting ticks in the **Red, Blue** or **Yellow** columns. More than one column may be ticked for each point.

c Add up the ticks to find out which route has the most advantages.

d Which route would you choose? The one with the most advantages would be best.

e Give **two** disadvantages of your chosen route.

f Describe the route you have chosen. Start with: 'The by-pass leaves the main road at ...'.

Considerations	Red route	Blue route	Yellow route
Is the shortest route			
Avoids all the built-up area			
Avoids best farmland			
Avoids steep slopes			
Avoids floodplain			
Avoids caravan park			
Needs fewest bridges			
Requires fewest trees to be cut down			
Avoids new housing estate			
Avoids sports park			
Total			

A

2 The **Yellow route** has some advantages but would be an unpopular choice for many reasons.

Work in pairs and suggest which of the people in **B** would be against the **Yellow route**. Give reasons for your answer.

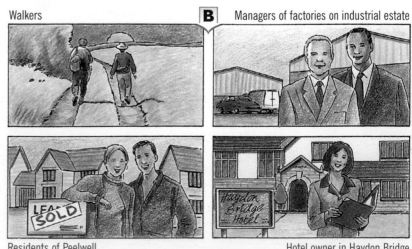

Walkers

Managers of factories on industrial estate

Residents of Peelwell

Hotel owner in Haydon Bridge

B

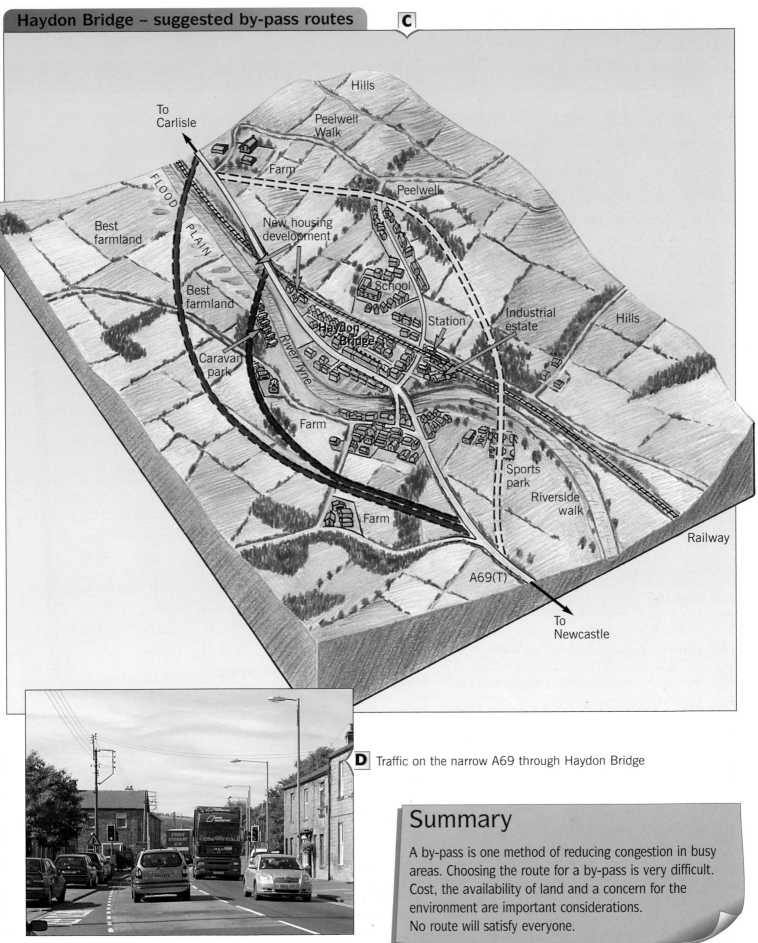

Haydon Bridge – suggested by-pass routes

C

Hills

Peelwell Walk

To Carlisle

Farm

Peelwell

FLOOD PLAIN

Best farmland

New housing development

Best farmland

School

Station

Industrial estate

Hills

Haydon Bridge

River Tyne

Caravan park

Farm

Sports park

Riverside walk

Farm

Railway

A69(T)

To Newcastle

D Traffic on the narrow A69 through Haydon Bridge

Summary

A by-pass is one method of reducing congestion in busy areas. Choosing the route for a by-pass is very difficult. Cost, the availability of land and a concern for the environment are important considerations.
No route will satisfy everyone.

How can a city street be improved?

Most main streets in UK towns and cities are busy and congested. They were built before people had cars and are unsuited to today's needs. Many streets are narrow and dangerous, and polluted by noise and exhaust fumes. There is no room to park and shopping can be an unpleasant experience. Buildings are often ugly and there is a lack of landscaping and open space. Overall, there is a poor **quality of environment**.

Drawing **A** shows a typical main street in a UK town with many of the problems just described. The people in the town want their street improved, and they have approached the local authority with their concerns. The authority agreed to produce a scheme that would:

1   reduce traffic congestion and pollution
2   make the area safer and more attractive
3   improve local shopping facilities.

Points are given for each feature.

For example:

◆ If a place is very attractive it will score 5 points.

◆ If it is ugly it will score 1 point.

◆ If it is in between it will score 2, 3 or 4 points.

The higher the number of points, the better the quality of environment.

QUALITY OF ENVIRONMENT SURVEY SHEET

	High quality				Low quality	
	5	4	3	2	1	
Attractive						Ugly
Quiet						Noisy
Tidy						Untidy
Safe						Dangerous
Few cars						Many cars
Easy movement						Congested
Good shopping						Poor shopping
Good parking						Poor parking
Open space						No open space
Like						Dislike

Place ....   Total .... out of 50

Drawing **B** shows the proposed improvement scheme. This must be discussed, and alterations suggested, before building can begin.

1 Make a list of the problems shown in drawing **A**.

2 Now look carefully at the improvement scheme in drawing **B**.

  a How has traffic congestion been reduced?

  b What has been done to improve safety?

  c What has been done to make the area more attractive?

  d How has shopping been improved?

3 **a** Make two copies of the survey sheet above.

  b Complete a survey for drawing **A**.

  c Tick the points you would give for each feature and add up the total number of points.

  d Complete a similar survey for drawing **B**.

4 Use the two surveys to measure the success of the improvement scheme. What features still need to be improved? Suggest what could be done to make these better.

5 The views of local people must be considered. For each person in **C**, say if they would be **for** or **against** the scheme. Give reasons for their views.

6 Should the scheme go ahead? Write a letter to the local authority giving your views and suggestions.

A The area before improvements

B Proposed improvement scheme

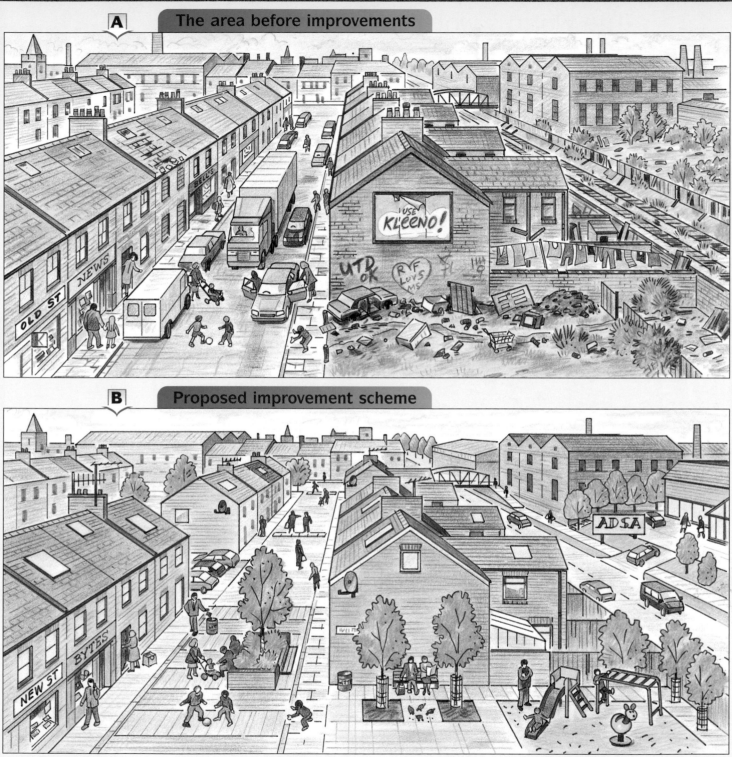

C

Mrs Briggs
A mother with two young children living on Main Street.

Mr Banks
Owner of a shop on Main Street. Often has to drive to his other shops in nearby towns.

Mr and Mrs Bell
Owners of a shop on Main Street which will be knocked down for parking.

The world's worst natural disaster?

What is this unit about?

This unit is about the Indian Ocean earthquake and tsunami of 2004 which was one of the world's most deadly and destructive natural disasters in recorded history.

In this unit you will learn about:

◆ the causes of the earthquake and tsunami
◆ the effects of the tsunami
◆ how the world responded to the disaster
◆ how the tsunami danger may be reduced.

◆ What is happening here?

◆ What are the likely dangers?

◆ What would you do if you were here?

A Phi Phi island, Thailand

Why is the Indian Ocean tsunami an important topic?

Many of you will have heard about the tsunami disaster and are aware of the enormous damage and loss of life that it caused. This unit will help you understand why it happened and what the effects were. Perhaps more important, it will also show how the world, and even you, can help in situations like this.

This unit can help you:

◆ understand a major world event

◆ be aware of the problems that a disaster causes

◆ appreciate the need for help in these situations

◆ know what you can do to help.

B | Patong Beach, Thailand

◆ What has happened here?

◆ If you were a survivor
  – what immediate help might you need
  – what problems might you have in the future
  – who do you think could help, and how?

◆ What could we in the UK do to help?

What caused the tsunami?

What happened?

On 26 December 2004, a massive undersea earthquake was recorded in the Indian Ocean off the north-western coast of Sumatra in Indonesia. The earthquake produced a **tsunami**, a series of huge waves that devastated the coastal areas of Indonesia, Sri Lanka, Thailand, southern India, and islands in the Indian Ocean, with waves up to 15 metres in height.

The earthquake was the fourth largest recorded in 100 years and was felt some 2,100 km (1,300 miles) away in India. The tsunami caused damage as far away as Somalia in East Africa, 4,500 km (2,800 miles) west of the earthquake's centre.

The most famous tsunami before that of 2004 followed the volcanic eruption of Krakatoa off the southern tip of Sumatra in 1883. The eruption caused the deaths of 36,000 people whilst the tsunami waves travelled several times around the world.

A

What is a tsunami?

♦ A **tsunami** is a giant wave or series of waves usually caused by an earthquake or volcanic eruption on the sea bed.

♦ As the tsunami travels through deep water, the wave may be less than a metre in height but it can reach speeds of up to 800 km/h (500 mph).

♦ When the wave nears the land it slows down and quickly increases to a height of up to 15 metres before crashing onto the coast.

The tsunami hits Sumatra **B**

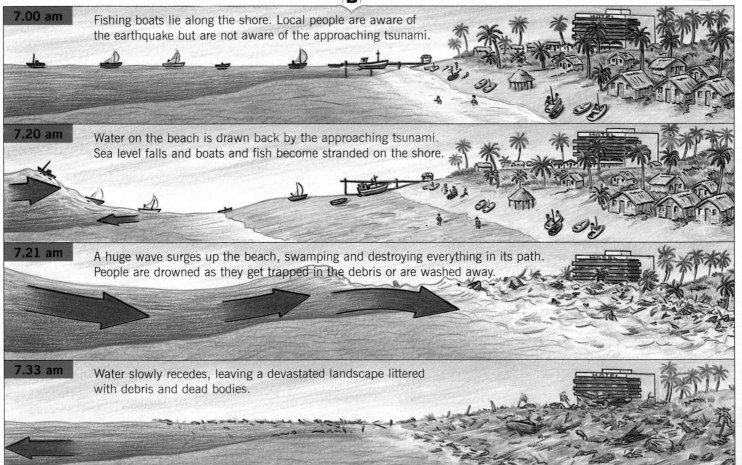

7.00 am Fishing boats lie along the shore. Local people are aware of the earthquake but are not aware of the approaching tsunami.

7.20 am Water on the beach is drawn back by the approaching tsunami. Sea level falls and boats and fish become stranded on the shore.

7.21 am A huge wave surges up the beach, swamping and destroying everything in its path. People are drowned as they get trapped in the debris or are washed away.

7.33 am Water slowly recedes, leaving a devastated landscape littered with debris and dead bodies.

What caused the earthquake and tsunami?

The outer crust of the earth is like a jigsaw. It is broken into huge pieces called **plates**. These plates move around very slowly. Where the plates meet they grind together and cause earthquakes. Volcanic eruptions also happen here.

The Indian Ocean earthquake occurred just west of Sumatra where the Indian and Eurasian plates meet.

As the plates moved towards each other they locked together and pressure built up. When the pressure was released there was an enormous earthquake.

The plate movement on 26 December 2004 was both sudden and massive. The sea above the earthquake was pushed upwards causing a tsunami. The tsunami waves then spread rapidly out from the earthquake in all directions.

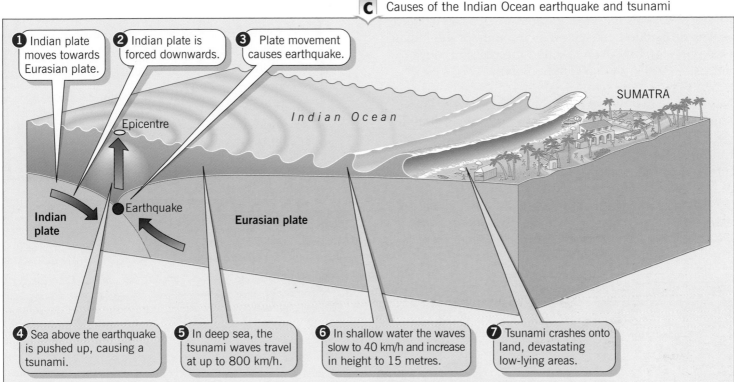

| **C** | Causes of the Indian Ocean earthquake and tsunami

1 Indian plate moves towards Eurasian plate.

2 Indian plate is forced downwards.

3 Plate movement causes earthquake.

Epicentre

Indian Ocean

SUMATRA

Indian plate

Earthquake

Eurasian plate

4 Sea above the earthquake is pushed up, causing a tsunami.

5 In deep sea, the tsunami waves travel at up to 800 km/h.

6 In shallow water the waves slow to 40 km/h and increase in height to 15 metres.

7 Tsunami crashes onto land, devastating low-lying areas.

Activities

1 Sort the statements below into the correct order to show the causes of the tsunami disaster.
- Sea movement causes tsunami.
- Plate movement causes earthquake.
- Waves crash onto shore.
- Sea above earthquake forced upwards.
- Plates move towards each other.
- Tsunami waves spread quickly outwards.

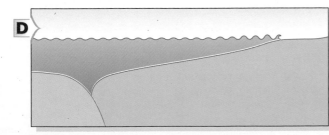

D

2 a Make a larger copy of cross-section **D**.
b Name the two plates, Sumatra and the Indian Ocean.
c Draw arrows to show plate movements.
d Label the earthquake position.
e Draw arrows to show the sea being pushed up and the movement of tsunami waves.
f Add a suitable title.

3 Describe the main features of a tsunami.

Summary

The Indian Ocean tsunami was produced by a massive earthquake off the coast of Sumatra. The earthquake was caused by two plates colliding with each other.

How did the tsunami affect different countries?

The Indian Ocean earthquake and tsunami of 2004 was one of the most deadly and destructive natural disasters in modern history.

The disaster began with the earthquake which destroyed towns and villages in northern Sumatra. This was followed by the tsunami which was to cause the most serious damage and loss of life. The tsunami waves devastated coastal areas throughout the Indian Ocean region and even reached the coast of Mexico almost 16,000 km (10,000 miles) away.

No-one will ever know the exact number of people killed as a result of the tsunami but it is believed to be over 310,000. As map **C** shows, Indonesia was worst-hit but some 11 other countries were also seriously affected by the tsunami waves.

So why was the death toll so high and the damage so extensive? The main reason is that the tsunami was extremely powerful and affected a very large area. These and some other reasons are shown in **A**.

A **Why were the effects so serious?**

◆ The earthquake was one of the most powerful ever recorded.
◆ Huge tsunami waves travelled thousands of kilometres across the Indian Ocean.
◆ Many coastal areas in the region are low-lying.
◆ Many areas are also densely populated.
◆ The disaster occurred during the main tourist season for visitors from Europe.
◆ There was no early warning system and no disaster plans in place.

B Tsunami wave hits Thailand

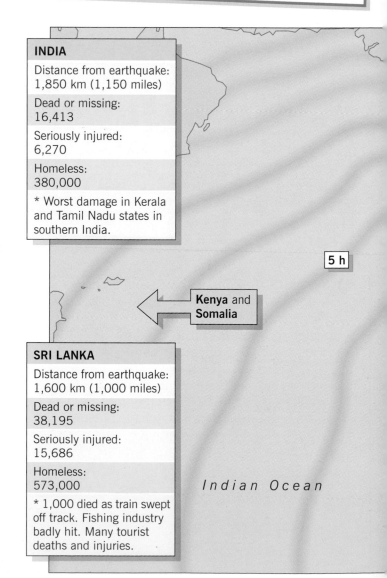

INDIA
Distance from earthquake: 1,850 km (1,150 miles)
Dead or missing: 16,413
Seriously injured: 6,270
Homeless: 380,000
* Worst damage in Kerala and Tamil Nadu states in southern India.

Kenya and Somalia

5 h

SRI LANKA
Distance from earthquake: 1,600 km (1,000 miles)
Dead or missing: 38,195
Seriously injured: 15,686
Homeless: 573,000
* 1,000 died as train swept off track. Fishing industry badly hit. Many tourist deaths and injuries.

Indian Ocean

Activities

1 Imagine that you are one of the people in photo **B** who survived the tsunami. Describe what happened, and your feelings at the time. Try to include these words:

- huge • crashing • roaring • panic
- frightened • run • escape • debris
- trapped • washed away • escape
- relief • friends • family

2 Complete a tsunami disaster FactFile using the table below. List the countries by the number of deaths. Give the highest first.

Country	Deaths	Injuries	Homeless	Distance from earthquake	Time for waves to reach

3 Draw a star diagram like **D** below, to explain the serious damage and loss of life. Write only four or five words in each box.

D

Why the effects were so serious

Summary

The devastation caused by the earthquake and tsunami was both serious and extensive. The damage and loss of life was worst in low-lying, densely populated areas.

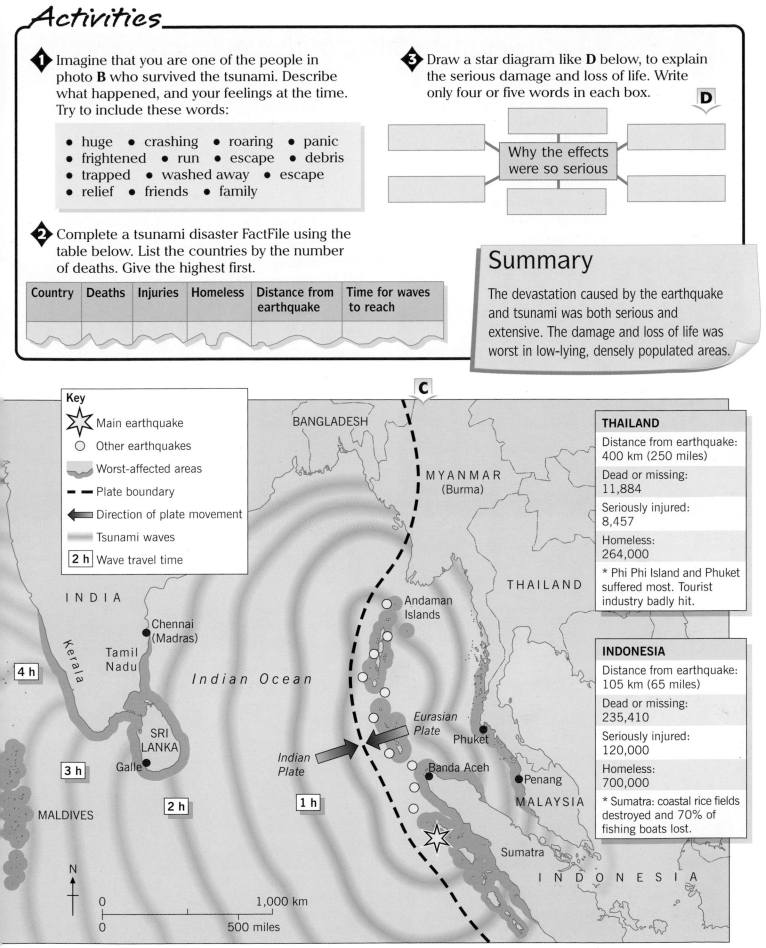

Key
- ⭐ Main earthquake
- ○ Other earthquakes
- Worst-affected areas
- – – Plate boundary
- ⬅ Direction of plate movement
- Tsunami waves
- 2 h Wave travel time

C

BANGLADESH

MYANMAR (Burma)

INDIA

Chennai (Madras)

Tamil Nadu

Kerala

Indian Ocean

Andaman Islands

THAILAND

SRI LANKA

Galle

Eurasian Plate

Phuket

Indian Plate

Banda Aceh

Penang

MALAYSIA

MALDIVES

4 h 3 h 2 h 1 h

Sumatra

INDONESIA

N

0   1,000 km
0   500 miles

THAILAND

Distance from earthquake: 400 km (250 miles)

Dead or missing: 11,884

Seriously injured: 8,457

Homeless: 264,000

* Phi Phi Island and Phuket suffered most. Tourist industry badly hit.

INDONESIA

Distance from earthquake: 105 km (65 miles)

Dead or missing: 235,410

Seriously injured: 120,000

Homeless: 700,000

* Sumatra: coastal rice fields destroyed and 70% of fishing boats lost.

What were the effects of the tsunami?

Reports on the death toll and damage caused by the earthquake and tsunami vary widely, due mainly to the enormous scale of the disaster. The figures shown in table **A** are estimates provided by the United Nations.

The effects of the disaster were both **short-term** and **long-term**. Some of these effects are shown in **B**.

Disaster toll

Dead or missing:	Over 310,000
Seriously injured:	Over 650,000
Homeless:	1.7 million
Jobs lost:	2 million
Reconstruction costs:	£6 billion
Aid promised:	£3.7 billion

A

B

1 Shelter
Most houses hit by the tsunami were simply washed away or damaged beyond repair. People were left with no shelter or place to live.

2 Schools and hospitals
Many were damaged and most had their contents totally destroyed. Many teachers, doctors and nurses were killed.

4 Jobs
Thousands of people lost their livelihood as the fishing industry was all but destroyed. Families were left with no money and no food.

5 Tourism
Many hotels and facilities were damaged. The tourist industry suffered badly. Income was reduced and local people lost their jobs.

The effects of the disaster were made worse by the fact that the countries affected are quite poor. These countries therefore do not have the money, organisation or technology to predict, plan for and cope with major natural disasters.

For example, there is no tsunami warning system in place for the region. If people had known that the wave was on its way, many thousands of lives may have been saved. Drawing **C** shows some of the problems.

Problems in poorer, less developed countries

C
- There is not the technology available to predict when a disaster might occur.
- Local rescue workers are poorly prepared and equipped.
- There are too few ambulances, hospital spaces, nurses and doctors.
- Buildings are often poorly constructed and easily damaged.
- There is a shortage of emergency clothing, shelter and medical supplies.

3 **Transport**
Coastal roads and railways were wrecked. Movement of emergency vehicles was limited. Travel afterwards became very difficult.

6 **Overcoming fear**
Many local people are frightened that another tsunami may hit the area. They worry about their future and way of life.

Activities

1 Look at the six photos in **B**. Choose the one that you think best shows the effects of the tsunami.

   a Describe the main features that it shows.

   b Imagine you were one of the survivors there. Describe what you would do and what your feelings would be.

2 Look at the photos again.

   a Describe two problems that are short-term.

   b Describe two problems that are long-term.

   c If you had been on holiday at a Thailand beach resort, would you stay or would you leave? Give reasons for your answer.

3 Describe the effects of the tsunami using the newspaper headlines shown below.

> Buildings Blasted by Giant Wave

> Transport links wrecked as tsunami hits coast

> Job worries for locals as industries washed away

> Future Bleak for Frightened Locals

4 Explain how natural disasters in poor countries usually do more damage than similar disasters in rich countries.
Use these headings:
- Disaster prediction
- Disaster preparation.

Summary

The Indian Ocean earthquake and tsunami occurred without warning and caused considerable damage and loss of life. Disasters affect poor countries more than rich countries.

How did the world help?

The Indian Ocean tsunami led to the biggest disaster relief effort ever known. Help came to the stricken countries from all around the world. It included food, shelter, medical supplies, machinery, expert assistance and money.

People, organisations and governments all gave help. Perhaps you gave something yourself or your school organised a money-raising event.

The following are some ways that help was given.

Organisations

International relief organisations which are used to dealing with natural disasters such as earthquakes and flooding tried to get blankets, tents, clean water, food and medical supplies in as fast as possible. This was not easy as local airports for the planes to land and roads to carry supplies to people had been destroyed.

People

Following appeals in newspapers and on television, ordinary people began phoning in or using the internet to give money for essential goods. People in the UK contributed over £100 million in only a few days and the Disaster Emergency Appeal (resource **E**) raised £300 million before closing down after two months.

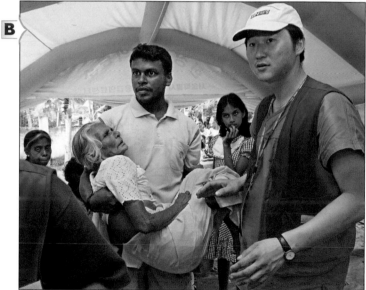

Many fund-raising events were also organised. These included international football and cricket matches and a concert in the Millennium Stadium in Cardiff.

Governments

Governments also gave help. They provided trained personnel, helicopters, lorries, heavy machinery and emergency supplies. They also gave a total of £3.7 billion to the disaster relief programme. Much of this money was put to long-term needs such as rebuilding schools and hospitals and restarting industries destroyed in the disaster.

Helping people caught up in a natural disaster is not easy. Providing the right kind of help at the right time can be a difficult job. People affected by the tsunami had different needs at different times. Priorities change from the time a disaster happens and during the weeks, months and even years afterwards.

Diagram **D** shows three different types of relief.

Usually people soon forget about a natural disaster, especially when these disasters take place a long way from home. Within a month of the tsunami, it was no longer in the world news. Within six months most government personnel had pulled out and the aid workers had left.

However, the relief operation was a success. It saved lives and helped countries recover from the enormous disaster that had hit them.

D

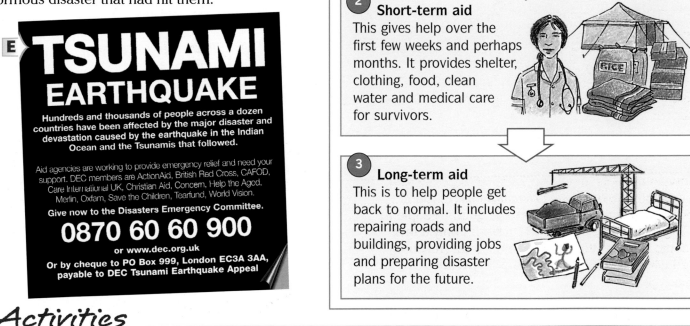

DISASTER!

1 Emergency relief
This gives immediate help. It provides rescue for those in danger, safety for survivors and emergency medical care and food for those in need.

2 Short-term aid
This gives help over the first few weeks and perhaps months. It provides shelter, clothing, food, clean water and medical care for survivors.

3 Long-term aid
This is to help people get back to normal. It includes repairing roads and buildings, providing jobs and preparing disaster plans for the future.

E

TSUNAMI EARTHQUAKE

Hundreds and thousands of people across a dozen countries have been affected by the major disaster and devastation caused by the earthquake in the Indian Ocean and the Tsunamis that followed.

Aid agencies are working to provide emergency relief and need your support. DEC members are ActionAid, British Red Cross, CAFOD, Care International UK, Concern, Christian Aid, Help the Aged, Merlin, Oxfam, Save the Children, Tearfund, World Vision.

Give now to the Disasters Emergency Committee.

0870 60 60 900

or www.dec.org.uk
Or by cheque to PO Box 999, London EC3A 3AA, payable to DEC Tsunami Earthquake Appeal

Activities

1 Draw a table like the one below and put the following examples of aid into the correct columns.

- Rescue teams • Tractors • Building equipment
- Shovels • Lorries • Rescue dogs
- Medical supplies • Food • Clean water
- Clothing • Cooking utensils • Blankets
- Tents • Education materials • First aid
- Expert help • Hospital equipment • Money

Emergency relief	Short-term aid	Long-term aid

2 Imagine that you were in Sri Lanka at the time of the disaster (photo **A**). Make a list of the things that you could do to help.

3 Think of something that you and your school could do to raise money for the disaster appeal. Design a poster to advertise your appeal. Resource **E** will help you.

Summary

The disaster relief programme was a worldwide effort. It brought both immediate and long-term help in the form of materials, money and expert assistance to the people and countries most in need.

How can the tsunami danger be reduced?

It is impossible to prevent earthquakes and tsunamis from happening but there is much that can be done to reduce the damage caused by them. Hawaii, for example, was hit by tsunamis in 1946 and 1964 following earthquakes in Chile and Alaska. In both cases, fewer than 100 people lost their lives and damage was small. This was due mainly to good preparation and an efficient early warning system.

Measures to reduce the effects of natural disasters such as earthquakes and tsunamis are usually in two parts. The first is to **predict** where and when the event will happen. The second is to **prepare** local people and the emergency services for the disaster. The following are some ways that this can be done.

1 Predict

Earthquakes

The accurate prediction of where and when an earthquake may happen is very difficult. Scientists use sensitive instruments and a variety of warning signs to help predict the next earthquake. Some are shown in **A**.

A
Earthquake warning signs

◆ Most earthquakes occur along plate boundaries where plates meet.

◆ Earthquakes are most likely after a long period without plate movement.

◆ There will be small foreshocks before the main earthquake which can be measured with a **seismometer**.

◆ Animals often act strangely. Snakes and rats may crawl out of their holes, and dogs howl.

Tsunamis

◆ Modern technology now allows us to identify and forecast tsunami strength and movement. A tsunami early warning system like the one in drawing **B** has been used successfully in the Pacific Ocean for many years.

◆ The Indian Ocean had no early warning system at the time of the disaster. The beginnings of a system have now been put in place and will be completed in stages over the next few years. A system for the North Atlantic is also planned.

◆ As with earthquakes, animals seem to sense the danger. In Sri Lanka elephants, leopards and other wildlife were seen to leave the danger areas before the tsunami hit. There were few reports of animal deaths.

B Early warning system

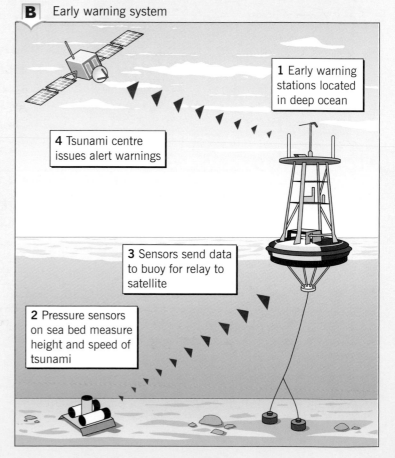

1 Early warning stations located in deep ocean

4 Tsunami centre issues alert warnings

3 Sensors send data to buoy for relay to satellite

2 Pressure sensors on sea bed measure height and speed of tsunami

2 Prepare

Sound preparation and planning can help limit the worst effects of an earthquake or tsunami. A good disaster plan should involve local authorities and emergency services as well as people living in the area.

Most earthquake deaths are due to fire and the collapse of buildings, so the best preparation is to have strict building regulations which are enforced.

Tsunamis are different. The best defence is a warning system and education so that people know to run for high ground if a warning is issued or if they feel an earthquake.

> **Now this is a fact …**
> A young British girl holidaying in Thailand was playing on the beach when the sea receded. She remembered from her geography lessons that this was a sign of an oncoming tsunami (page 90). She warned people and hundreds escaped almost certain death.

C

Activities

1. Write down three signs that an earthquake may be about to happen.

2. **a** How could a tsunami early warning system save lives?

   b Would early warning always save lives? Give reasons for your answer.

3. Write out the sentence beginnings in drawing **D** and complete them with the correct endings from the following list:

   … follow the rules for safe buildings.
   … practise what to do in an emergency.
   … help prepare disaster and evacuation plans.
   … organise emergency supplies of food and water.
   … be prepared to give disaster advice.
   … run for high ground in an emergency.

4. List the different ways that could be used to inform people about earthquakes and tsunamis.

D

Rescue services should …

Radio and TV stations should …

Local authorities must …

Builders must …

Every month, schools must …

EMERGENCY PLANNING GROUP

Everyone should …

Summary

It is impossible to prevent earthquakes and tsunamis from happening. Good preparation and planning can help save lives and reduce damage to property.

In this enquiry you should imagine that you work for the Disaster Emergency Agency in Sri Lanka. Funds for the agency's work are raised in the UK.

Your job is to decide how best to use the funds to provide relief for those affected by the tsunami disaster. Your area of responsibility is south-west Sri Lanka, around Galle. The map on pages 92–93 shows where this is. The photos on pages 89, 94, 95, 96 and 99 will also help you.

The huge waves of the Indian Ocean tsunami hit your region at 7.14 am in the morning and devastate the coastal areas. Thousands of people are reported dead, injured or missing. The damage is enormous.

You hold a crisis meeting with your team of experts to discuss possible schemes and costs. You have £500,000 to spend but must work fast.

It will be best if you can work with a partner or in a small group. You will then be able to share views and discuss ideas with each other.

How can we help people affected by the tsunami disaster?

1 a Look carefully at the schemes on the opposite page and discuss which are the most important.

b Sort the schemes under these headings:
- Immediate and short-term
- Long-term.

2 a Copy the table (right) and list all the schemes.

b Discuss which you will use and how much you will spend on each one.

c Complete your table.

> **Remember ...**
> ✔ You do not have to spend the full amount for each scheme but, if you don't, the scheme will be less effective.
> ✔ The total cost must be no more than £500,000.
> ✔ Your decisions must reflect the aims of the Disaster Emergency Agency.

Disaster Emergency Agency

Aims
- ◆ To provide immediate help and save lives
- ◆ To care for those in need and provide long-term support
- ◆ To support appropriate and sustainable long-term solutions to local problems
- ◆ To support projects that enable local people to help themselves.

Scheme	Total cost	Money spent
Feeding scheme	£50,000	

3 a Make a list of your choices and their costs in order of importance.

b Give reasons for your choice of the first two.

4 Write a short report for the Project Director giving details of your decisions. Once again, look carefully at the aims of your agency.

Feeding scheme
Provide food and water for immediate use and for up to three months afterwards.
Cost £50,000

Emergency warning system
Contribute towards disaster plan and Indian Ocean early warning system.
Cost £25,000

Home building scheme
Build new housing for those who lost their homes.
Cost £50,000

Hospital reconstruction
Repair and re-equip all medical centres including hospitals.
Cost £200,000

Job provision
Rebuild livelihoods by restarting industries such as fishing, farming and tourism.
Cost £80,000

Shelter and warmth
Provide tents, clothing and blankets for survivors made homeless.
Cost £50,000

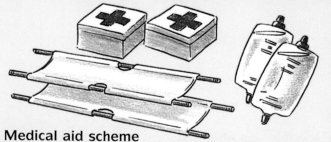

Medical aid scheme
Provide medical equipment and first aid packs for emergency use.
Cost £30,000

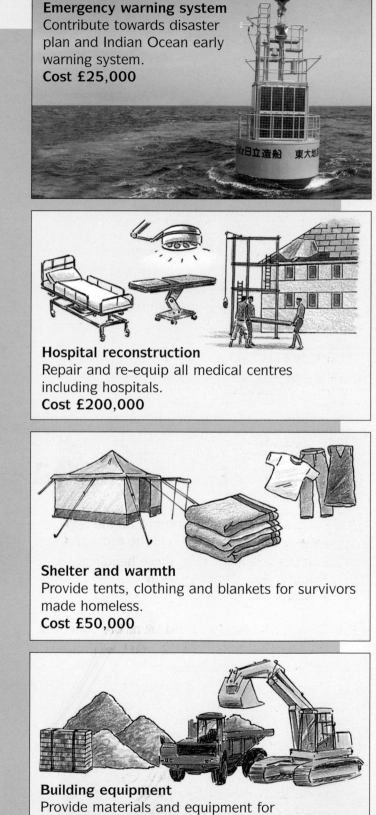

Building equipment
Provide materials and equipment for rebuilding programme.
Cost £80,000

What is the UK like?

What is this unit about?

There are three main parts to this unit. The first looks at the UK's main features and the second at regional differences. The final part is concerned with your own local area.

In this unit you will learn about:

◆ the UK's physical and human geography

◆ where people in the UK come from

◆ the wealth of the country and its regional differences

◆ what your local area is like.

Why is learning about the UK important?

Geography is about people and places. Learning about the UK helps us to understand the country and makes it a more interesting place in which to live. It will also help you make more sense of events in the news and help you develop your own views and opinions about both local and national issues.

The unit can also help you to:

◆ learn about the UK's main features

◆ learn about and appreciate your surroundings

◆ understand where people in the UK come from and how their backgrounds and beliefs vary

◆ develop an interest in your own local area.

A Leeds

For each of the four photos, make a list of words to describe it.

◆ For places A and B, which
– would you most like to visit
– is most different from where you live?

Give reasons for your answers.

B **Yorkshire Dales**

D **Near Birmingham**

Newcastle

C

103

Where is the UK?

The world can be divided into seven large landmasses and several large sea areas. The land masses, which cover 30 per cent of the earth's surface, are called **continents**. The five largest sea areas are known as **oceans**. The continents and oceans are shown on map **A**.

The **United Kingdom**, often referred to as the **UK**, is located to the north-west of the continent of Europe. Before 2004 it was one of 15 countries that had joined together to form a group known as the **European Union** – the **EU** for short. These countries, shown on map **B**, are mainly in western Europe. In 2004, the EU was joined by 10 more countries, mainly from eastern Europe.

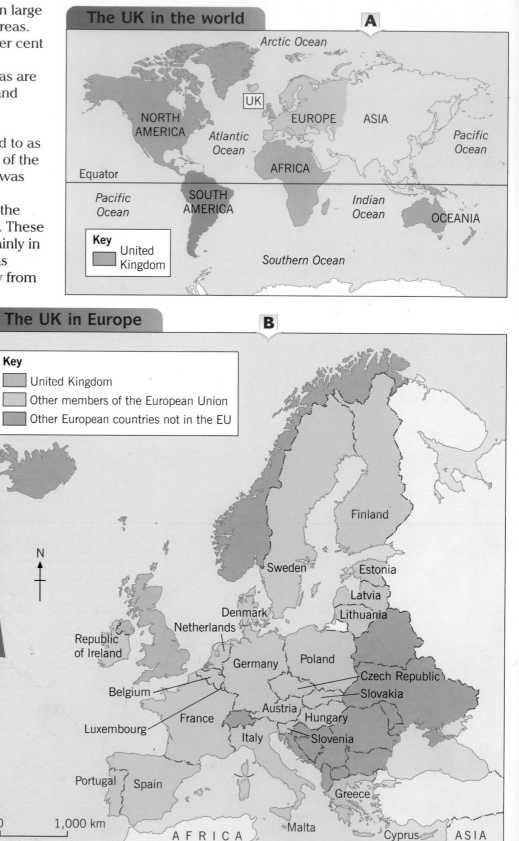

The UK in the world **A**

Arctic Ocean
UK
NORTH AMERICA
Atlantic Ocean
EUROPE
ASIA
Pacific Ocean
Equator
AFRICA
Pacific Ocean
SOUTH AMERICA
Indian Ocean
OCEANIA
Southern Ocean

Key
United Kingdom

The UK in Europe **B**

Key
United Kingdom
Other members of the European Union
Other European countries not in the EU

Finland
Sweden
Estonia
Latvia
Lithuania
Denmark
Netherlands
Republic of Ireland
Germany
Poland
Czech Republic
Slovakia
Belgium
Austria
Hungary
Luxembourg
France
Slovenia
Italy
Portugal
Spain
Greece
Malta
Cyprus
AFRICA
ASIA
N
0   1,000 km

The EU flag

What is the UK?

Maps **C**, **D** and **E** explain the differences in meaning between the **British Isles**, **Great Britain** and the **United Kingdom**.

Each of the four countries that form the UK has a certain amount of self-rule. This means that each is able to develop its own distinct characteristics. These characteristics include language and culture together with each country's own legal and education system.

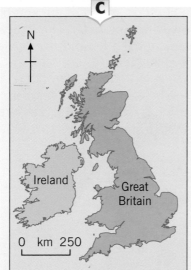

C

The British Isles consist of two large islands. These islands are called Great Britain and Ireland.

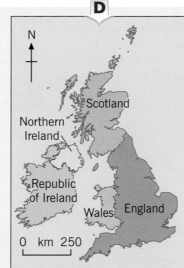

D

Great Britain, the largest island, consists of three countries – England, Wales and Scotland. Ireland is divided into two countries – Northern Ireland and the Republic of Ireland.

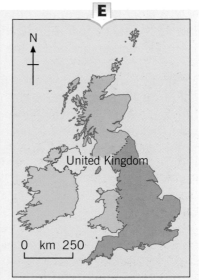

E

The United Kingdom consists of the four countries of England, Wales, Scotland and Northern Ireland. The Republic of Ireland is an independent country.

Activities

1 What is the difference between:
   a the British Isles
   b the United Kingdom and
   c Great Britain?

2 Using the information in table **F**, what is the:
   • total population
   • area of
   a the British Isles   **b** Great Britain   **c** the UK?

3 Make a larger copy of table **G**. Sort the following places into columns. Some may go into more

   Africa, Antarctica, Asia, Belgium, Denmark, England, Europe, Finland, France, Germany, Greece, Italy, Luxembourg, Netherlands, North America, Northern Ireland, Oceania, Portugal, Republic of Ireland, Scotland, South America, Spain, Sweden, UK, Wales.

F

	Area (km²)	Population (thousands)
England	130,423	49,856
Northern Ireland	14,160	1,703
Rep. of Ireland	70,280	3,825
Scotland	78,133	5,057
Wales	20,766	2,938

G

Continents	EU countries	UK

Summary

The UK consists of four countries. It is located in north-west Europe and is a member of the European Union.

105

What are the UK's main physical features?

Map **A** is a **physical map** of the UK. It shows the main sea areas, the largest islands, the longest rivers, the largest mountain areas and the highest mountains.

Most of the north and west of the UK is mountainous (photo **B**). This is because the rocks found here are old and tend to be resistant to erosion. The area has, over millions of years, been slowly worn away to leave rugged mountain peaks and deep, often lake-filled, valleys.

In contrast, much of the south and east of the UK is low-lying (photo **C**). The typical scenery is one of wide and flat-floored river valleys separated by areas of rolling hills.

Nowhere in the UK is far from the sea. The coastline, which is over 11,000 km in length, has scenery which varies from high cliffs to low-lying sand dunes.

For such a small country, there are also considerable differences in weather and climate between different places in the UK. Temperatures are usually higher in the south of England than they are in the north of Scotland (map **D**). Rainfall, which can fall at any time of the year, is often greater in the west of Britain than it is in the east (map **E**).

The UK is said to have a **variable** climate because the weather often changes from day to day (if you have made daily recordings of the weather you will have

noticed this for yourself). The climate is also said to be **temperate**, with warm summers, mild winters and some rain falling throughout the year. It is unusual for the UK's weather to be either too hot, too cold, too dry or too wet. However, although the UK usually avoids such weather hazards as drought, heat waves or extreme cold, the frequency and severity of flooding and storms does seem to have increased in recent years. The UK's weather is described in more detail in Chapter 2.

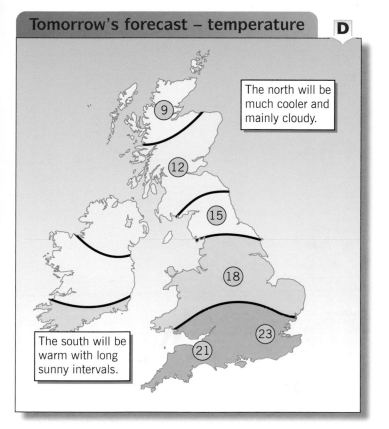

Tomorrow's forecast – temperature **D**

The north will be much cooler and mainly cloudy.

9

12

15

18

23

21

The south will be warm with long sunny intervals.

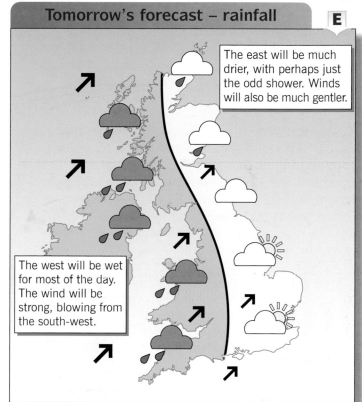

Tomorrow's forecast – rainfall **E**

The east will be much drier, with perhaps just the odd shower. Winds will also be much gentler.

The west will be wet for most of the day. The wind will be strong, blowing from the south-west.

Activities

1 a Look at photo **B** and maps **D** and **E**.
- Where in the UK was the photo taken?
- Describe the scenery in the photo.
- Describe the climate of the area.
- Would you like to live there?

Give reasons for your answer.

b Look at photo **C** and maps **D** and **E**. Answer the same questions as you did in part **a**.

2 a In what ways is the area near to where you live:
- similar to
- different from

those shown in photos **B** and **C**?

b According to the information shown on maps **D** and **E**, what is the weather forecast for your local area?

Summary

Both the scenery and the climate of the UK are varied. Usually the scenery is very attractive while the climate is temperate and without extremes.

How is the UK divided up?

The United Kingdom is divided up into many different pieces, just like a jigsaw. Indeed, as we shall see, there are many different jigsaws, each one smaller than the next. They include countries, regions, counties, districts and parishes. The divisions are made to help run the country efficiently. The maps on these two pages show some of the pieces.

Map **A** shows the United Kingdom, which is a country made up of England, Wales, Scotland and Northern Ireland. This country is governed by the Parliament at Westminster in London which is attended by members from all four countries.

Within this system, however, each country has a certain amount of self-rule and is able to develop its own distinct characteristics. Scotland, for example, has its own education system and legal system and its own forms of local government. Wales has its own language and some control over its own affairs. Northern Ireland has separate laws and a distinct system of education.

Map **B** shows what are called the standard or administrative regions of the UK. These are used for collecting information and statistics and are useful for comparing different parts of the country. The statistics may be accessed on the government's **National Statistics** website. Suggestions on how to use this may be found on pages 120 and 121.

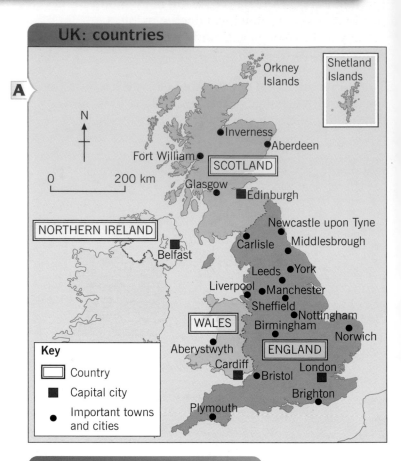

UK: countries

A

Key
☐ Country
■ Capital city
● Important towns and cities

UK: administrative regions

B

After the regions we have the counties. These are shown on map **C** and are important pieces in another jigsaw. They are used for local government and are often the part of the country that local people most identify with. They usually have a character of their own and people are proud of the county they come from.

But that's not all! These counties are further divided into districts and parishes. There is not the space in this book to show them but you might like to find out about these regions in the area where you live.

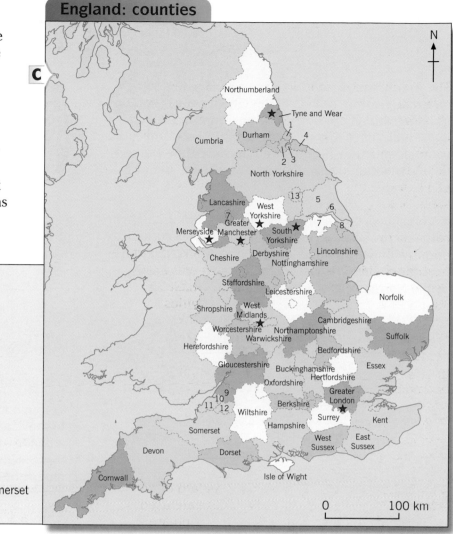

England: counties

C

Key

★ Metropolitan counties

Unitary authorities:
1 Hartlepool
2 Stockton-on-Tees
3 Middlesbrough
4 Redcar and Cleveland
5 East Riding of Yorkshire
6 Hull
7 North Lincolnshire
8 North East Lincolnshire
9 South Gloucestershire
10 Bristol
11 North Somerset
12 Bath and North East Somerset
13 York

Activities

❶ Look at maps **A**, **B** and **C**.
Name a city and three counties in:

a the East region
b the South East
c the South West
d the North West
e the East Midlands.

❷ In which administrative region is:

a Aberdeen   **b** Aberystwyth   **c** Birmingham
d Leeds   **e** Belfast?

❸ Use the headings in drawing **D** to describe where you live. The maps on these two pages and the one on the back cover will help you.

❹ What are the characteristics of where you live? Make a list or write a paragraph to describe them.

The place where I live ...

D

◆ Country ...
◆ Region ...
◆ County ...
◆ Nearest town ...
◆ District ...
◆ Parish ...
◆ Other features ...
(Try to give at least four.)

Summary

The United Kingdom is divided into many different areas. They include countries, administrative regions and counties. Each has a part to play in the running of the UK.

Where do people in the UK come from?

The movement of people from one place to another is called **migration**, and a person who moves is known as a **migrant**. People, whether in groups or as individuals, have always moved. Table **A** explains how some people may have moved because they wanted to, and others because they had no choice and just had to move.

Map **B** shows the many early migrants into Britain who came as **invaders**. While some may have returned home with their plunder, others became **settlers** and remained here. The last major invasion of Britain was by the Normans in 1066.

A

Voluntary migration is when people choose to move. This may be:	
• to improve their standard of living	– for more or better-paid jobs
• to improve their quality of life	– retiring to a better climate
	– living/working in a more pleasant environment or with better services.
Forced migration is when people have little or no choice but to move. This may result from:	
• natural disasters	– earthquakes, volcanic eruptions or floods
• economic, social or political pressures	– war, famine, religious or political persecution.

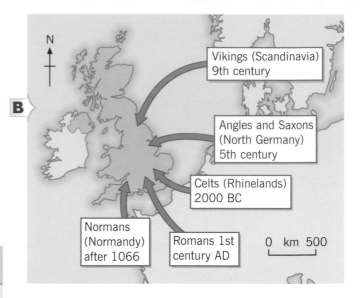

B

Vikings (Scandinavia) 9th century

Angles and Saxons (North Germany) 5th century

Celts (Rhinelands) 2000 BC

Normans (Normandy) after 1066

Romans 1st century AD

0   km 500

The British population

Throughout the twentieth century, people continued to move into and out of the UK. People who move into a country like the UK are called **immigrants**, while those who move out of a country are known as **emigrants**. Diagram **C** shows some of the recent immigrant groups into the UK, and gives reasons why they came to settle here.

The effect of migration into the UK has been considerable. It has caused an increase in numbers and changed the mix of people in the country (graph **D**). It has produced a **multicultural society** where people of different ethnic groups, languages, religions and cultures live and work together (photo **E**). While this mix can cause problems, most people agree that it has added variety and created benefits.

C

A large number of Chinese also now live in the UK.

My grandparents came from the West Indies in the 1950s. They were encouraged to come by the British government as the UK had a serious labour shortage after the Second World War.

Many immigrants came from Eastern Europe in the 1940s and 1950s to escape war and religious and political persecution.

Many immigrants to Britain have come from India, Pakistan and Bangladesh. They came to Britain to seek work and a better education, to start their own businesses and to improve their standard of living.

Many generations of Irish have come to Britain, mainly to find jobs.

Recent immigrants into the UK

Diagram **F** shows that, during the 1990s, three out of every five immigrants into the UK came from the EU. This was mainly due to the ease of moving from one EU country to another. By 2000, however, certain people in the UK were becoming increasingly worried about three types of immigrant:

◆ **Refugees** who come to Britain because they might be tortured, imprisoned or killed in their home country. This persecution may be due to their ethnic grouping or their religious or political beliefs, e.g. people from the former Yugoslavia.

◆ **Asylum seekers** are people who live in danger in their own country and want to move to another country where they will be safe.

◆ **Illegal immigrants** who try to enter the UK illegally, e.g. the 56 Chinese who suffocated in a lorry as it came across the Channel.

D Ethnic groups in the UK, 2001

Mixed 1.1%
Black/Black British 2.0%
Chinese 0.4%
Others 0.4%
Indian, Pakistani & Bangladeshi 4.0%
White 92.1%

E

F Source of the most recent migrants into the UK

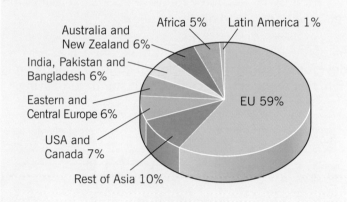

Australia and New Zealand 6%
Africa 5%
Latin America 1%
India, Pakistan and Bangladesh 6%
Eastern and Central Europe 6%
USA and Canada 7%
Rest of Asia 10%
EU 59%

Activities

1 a Make a list of where everyone in your class was born. Plot the information on a map of the UK or, if many were born outside the UK, on a map of the world.

b Draw two types of graph to show where people in your class were born. Pages 16 and 17 will help you decide which types of graph to use. You may be able to draw your graphs on a computer. The graphs might show those born in each of the following places:
  ● within 10 km of your school
  ● over 10 km from the school but within the same region or county
  ● elsewhere in the UK
  ● in one of the other 24 EU countries
  ● elsewhere in the world.

2 Your class should divide into four or five groups. Each group should then investigate one of the groups of people who have moved into the UK since 1900, e.g. West Indians, or refugees from the former Yugoslavia. The questions on clipboard **G** should help you with your investigation.

G

a Name of a group of immigrants.
b Where did they come from?
c Why did they leave their home country?
d When did they come to the UK?
e Why did they come to the UK?
f Once in the UK, where did most of them settle?
g Why did they settle there?
h What contribution have they made to the community in which they live?

Summary

The UK is made up of people from many different countries. Different ethnic groups and differences in language, religion and culture have given the UK a multicultural society.

How well off is the UK?

The UK is one of the wealthiest countries in the world. Its success is due mainly to the growth of industry, and trade with other countries. The UK's most important trading partners are countries in the **European Union (EU)**.

Most workers in the UK earn good wages. This is largely because the majority of them work in manufacturing and tertiary industries which are generally well paid (diagram **A**). This has enabled them to enjoy a high **standard of living** and a good **quality of life**. They have money to spend on holidays, food and education. They can also afford to buy their own cars and expensive goods like television sets, computers and digital cameras.

All of these things have helped make the UK a **developed country**. A developed country is one that is rich, has many services and a high standard of living.

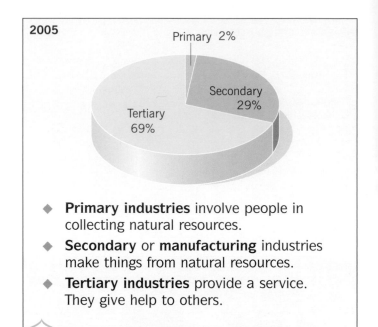

2005

Primary 2%
Secondary 29%
Tertiary 69%

◆ **Primary industries** involve people in collecting natural resources.

◆ **Secondary** or **manufacturing** industries make things from natural resources.

◆ **Tertiary industries** provide a service. They give help to others.

A UK employment structure

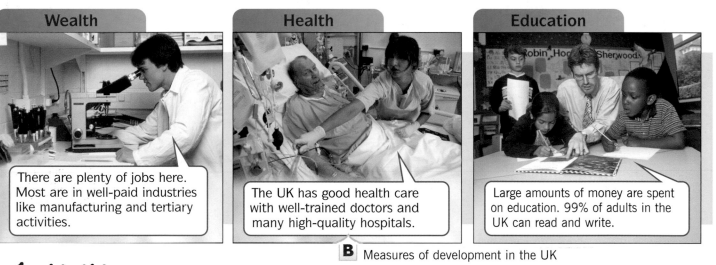

Wealth

There are plenty of jobs here. Most are in well-paid industries like manufacturing and tertiary activities.

Health

The UK has good health care with well-trained doctors and many high-quality hospitals.

Education

Large amounts of money are spent on education. 99% of adults in the UK can read and write.

B Measures of development in the UK

Activities

1 Complete these sentences. The Glossary on pages 142–144 will help you.

  a Standard of living is a measure of ...

  b Quality of life is a measure of ...

  c A developed country is ...

2 a List the jobs from drawing **C** under the headings:
   • Primary • Secondary • Tertiary.

  b Add at least two more jobs under each heading.

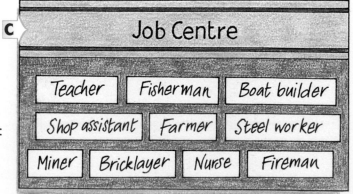

C

Job Centre

Teacher	Fisherman	Boat builder
Shop assistant	Farmer	Steel worker
Miner	Bricklayer	Nurse	Fireman

Although the UK is a developed country, wealth and high standards of living are not shared equally between everyone. All cities, for example, have some areas that are rich and some that are poor.

Even across the country, there are differences in wealth and standards of living. Look at map **D** which shows the average weekly earnings in the UK. Notice the difference between the north and south.

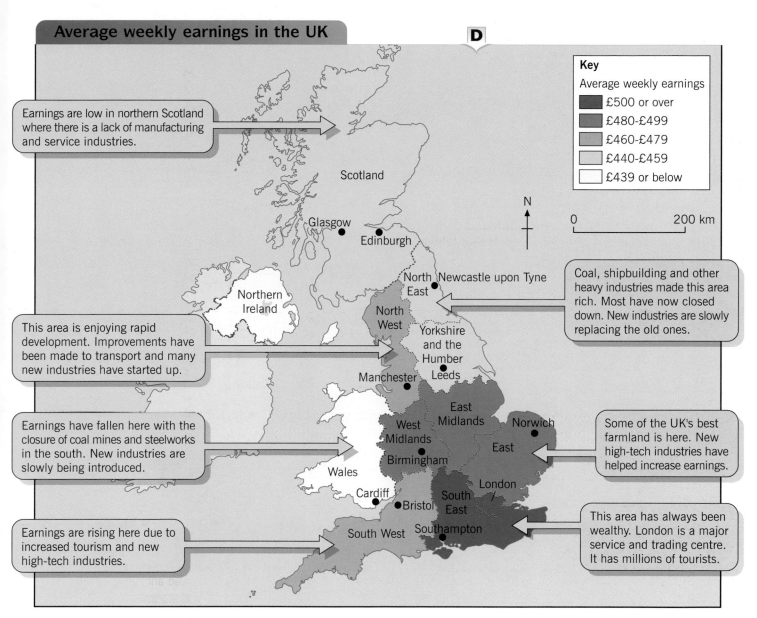

Average weekly earnings in the UK

D

Earnings are low in northern Scotland where there is a lack of manufacturing and service industries.

This area is enjoying rapid development. Improvements have been made to transport and many new industries have started up.

Earnings have fallen here with the closure of coal mines and steelworks in the south. New industries are slowly being introduced.

Earnings are rising here due to increased tourism and new high-tech industries.

Coal, shipbuilding and other heavy industries made this area rich. Most have now closed down. New industries are slowly replacing the old ones.

Some of the UK's best farmland is here. New high-tech industries have helped increase earnings.

This area has always been wealthy. London is a major service and trading centre. It has millions of tourists.

Key
Average weekly earnings
£500 or over
£480–£499
£460–£479
£440–£459
£439 or below

N

0   200 km

Scotland
Glasgow
Edinburgh
Northern Ireland
North East   Newcastle upon Tyne
North West
Yorkshire and the Humber
Manchester   Leeds
East Midlands
West Midlands   Norwich
Birmingham   East
Wales
Cardiff   London
Bristol   South East
South West   Southampton

3 Use map **D** to answer this activity.
  a The average earnings in the North East are …
  b The wealthiest region is …
  c The regions with the lowest earnings are …
  d The average earnings in my region are …
  e Five things that help make a region rich are …

4 Describe the pattern of wealth in the UK.

5 Think about the lifestyle that you live. Give at least six ways that your lifestyle can be described as typical of someone who lives in a developed country like the UK.

Summary

The UK is one of the most developed countries in the world. Development is not spread evenly. Some people still have a poor standard of living.

What are the regional differences in the UK?

The map on this page is a **satellite image** of the UK. A satellite image is a photo that is taken from space and sent back to Earth. Many satellite images have 'false colours'. This means that colours may be changed slightly to highlight, or enhance, certain characteristics, e.g. the extent of flooding, a weather system (page 30); or the effect of deforestation.

On this satellite image:

◆ dense forest and areas of lush vegetation show up as bright green
◆ highland and other places that lack vegetation appear as various shades of brown
◆ rivers, lakes and sea are dark blue, and
◆ urban areas are pink.

We have already seen on pages 108–109 that the UK can be divided into several regions. In map **B**, the number of regions has, for simplicity, been reduced to six. The main characteristics of relief, climate, settlement and economic activities for each of these regions, appear on map **A**. You will notice that each of these regions has:

◆ some similarities with other regions
◆ some differences from other regions.

A

Wales
• Mainly highland with mountains in the north.
• Heavy rainfall all year, especially in winter.
• Warm summers and mild winters in low-lying areas.
• Colder, with more wind, in highland areas.
• Most people live in the south, elsewhere is mainly rural.
• Secondary and service activities in the south, mainly primary elsewhere.

South and west of England
• Many hills and moorland areas.
• Heavy rain all year, especially in winter.
• Warm summers and very mild winters.
• Mainly rural with a few urban centres.
• Mainly primary and service activities.

Scotland
- Mainly mountainous.
- Central area, with the rivers Forth and Clyde, is lower.
- Heavy rainfall all year, especially in the west.
- Highest areas get snow and strong winds in winter.
- Cool summers, mild winters.
- Most people live in central areas, elsewhere is mainly rural.
- Secondary and service activities in central areas, mainly primary with some service activities in other areas.

North and east of England
- Low-lying apart from the Pennines in the west.
- Light rainfall all year.
- Cool summers, cold winters.
- Many large urban areas.
- Decline in primary and secondary activities, increase in service activities.

North and west of England
- Includes the highlands of the Lake District and Pennines.
- Heavy rainfall all year, especially in highland areas and in winter.
- Cool summers, mild winters.
- Mainly urban to the south and west of the region, mainly rural to the north and east.
- Secondary activities in the south, primary activities in the north and east, services more widespread.

South and east of England
- Mainly low-lying with gentle hills.
- Several long rivers, such as the Thames.
- Some rain all year, most in summer.
- Cold winters, warm summers.
- Both urban and rural.
- Many service activities, few primary or secondary jobs.

Activities

B

N

Scotland

0 km 200

North and East England

North and West England

Wales

South and East England

South and West England

1 Match the following regions with the best descriptions:

Scotland — is mainly flat and low-lying.

Wales — is mountainous with a central lowland.

North and west England — is the mildest region.

North and east England — includes the highest land in England.

South and east England — is hilly and has cold winters.

South and west England — has mountains in the north.

2 a Make a FactFile to show the main characteristics of your local area or region. Use the headings shown in FactFile **C**.

b Make another FactFile using the same headings. This time, choose a different region to the one in which you live. Say in what ways this region is similar to or is different from yours.

C

FactFile
- Relief ...
- Climate ...
- Settlement ...
- Economic activities ...

Summary
The UK can be divided into several regions. Each region has its own characteristics of climate, relief, settlement and economic activities.

How can maps help us describe the local area?

You can learn and understand a lot of geography by using large-scale maps of the area surrounding your home and school, and by doing your own local fieldwork. Unfortunately it is impossible to provide local maps or to ask questions specific to the local area of every school using this textbook. However, as most people of your age in the UK live or go to school in towns or cities, you may well recognise the types of area shown in maps **A** and **B**. You could then use your local map and adapt the activities opposite to fit that.

Map **A** shows the area around a school near to the centre of a large city. Map **B** shows the area around a school that could be either on the edge of a large city or within a small town.

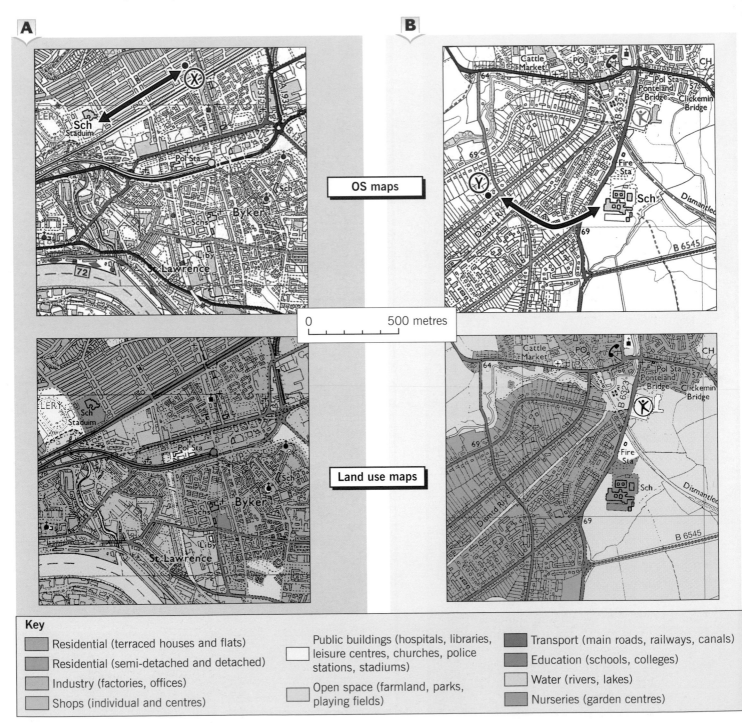

A

B

OS maps

0   500 metres

Land use maps

Key

◻ Residential (terraced houses and flats)

◻ Residential (semi-detached and detached)

◻ Industry (factories, offices)

◻ Shops (individual and centres)

◻ Public buildings (hospitals, libraries, leisure centres, churches, police stations, stadiums)

◻ Open space (farmland, parks, playing fields)

◻ Transport (main roads, railways, canals)

◻ Education (schools, colleges)

◻ Water (rivers, lakes)

◻ Nurseries (garden centres)

Activities

1
a What are the two main types of land use on map **A**?

b Name three other types of land use on map **A**.

c What are the two main types of land use on map **B**?

d Name three other types of land use on map **B**.

e Suggest two reasons for the differences in land use between maps **A** and **B**.

2 Each map shows the route taken by a pupil from their home to their school.

a Make a larger copy of table **C**.

b Complete the table using words and terms from the box below. Pages 70 and 71 will help you with your answer.

- detached and semi-detached • terraced
- built with brick • built with stone
- over 100 years old • less than 30 years old
- curving roads • straight roads
- inner city zone • outer suburbs zone

C

Features	House X on map A	House Y on map B
Type of house		
Building material		
When built (age of house)		
Shape of roads		
Zone within a city		

3 Now describe your own route from home to school. Use the headings and questions on clipboard **D**.

D

Buildings
What do they look like? How old are they? What materials are used?

Shops
How many are there? What type are they? How busy are they?

Roads
How many main roads do you cross? How busy and noisy are they?

Places of work
Are there any factories, offices or other places of work? What are they like?

Environment
Are there any areas of parkland or open countryside? Is the area tidy or messy?

4 Now look at a map of your local area.

a Mark on it your home and your school.

b Mark the route you usually take from your home to your school.

c Colour in and name the main types of land use through which you pass. (Use the terms given in the map key on the opposite page.)

d Name, colour and add to your key any other important types of land use on your route.

Summary

You can learn about your local area by using maps, the internet or by fieldwork.

What is it like where you live?

What is your environment like?

The local environment is the place in which we live, work and play. The local environment may still be very pleasant, as in photo **A**, or it may have been damaged or spoilt by people living in it before you, as in photo **B**. It may still be being spoilt today. To prevent the environment from being spoilt further, or to try to improve it if it has already been damaged, it needs to be protected and carefully managed and looked after.

A

B

You can use an **environmental survey** to measure the quality of your local area. This survey will help you to identify any damage or problems. It can then be used to help solve the problems and, hopefully, to improve you local surroundings.

Diagram **C** shows a typical environmental survey sheet. It suggests 10 features that might make an environment either pleasant or spoilt. Activity 1 suggests that you give points to each feature – the higher the points you award, the better you think it is. You may be able to add other features to the list or substitute terms to suit your local area.

C

QUALITY OF ENVIRONMENT SURVEY SHEET

	High quality				Low quality	
	5	4	3	2	1	
Attractive						Ugly
Quiet						Noisy
Tidy						Untidy
Safe						Dangerous
Few cars						Many cars
Easy movement						Congested
Good shopping						Poor shopping
Good parking						Poor parking
Open space						No open space
Like						Dislike

Place .... Total .... out of 50

Activities

1 **a** Put one tick ✔ in each horizontal row in **C**.

b Add up the total points you have awarded and then multiply by two to give a percentage.

c Complete an environmental survey sheet for at least one other place in your local town.

d In which of the environments would you prefer to live? Give **three** reasons why you would prefer to live there. Give **three** reasons why you would not want to live in the place, or places, that you rejected.

✔ 5 if the place is **attractive**
✔ 4 if the place is **fairly attractive**
✔ 3 if the place is **neither attractive nor ugly**
✔ 2 if the place is **quite ugly**
✔ 1 if the place is **ugly**
The higher the number, the better the quality.

What is your quality of life like?

By **quality of life** we mean how content we are with our lives and with the environment in which we live. Contentment can include many different things. For example, it could be:

- how happy we are
- how well off we are
- how secure we feel
- how many friends we have
- how many things there are for us to do
- how we like living and working in the local area.

Quality of life differs greatly between countries, within countries, within towns and even from house to house. It may also vary from family to family and from person to person. How content are you with your quality of life?

Diagram **D** shows 20 features that can help you measure more accurately whether your quality of life is good, average or poor. Once again each of the features can be given a score of 1 to 5. If each feature were given a maximum score of 5, the total would be 100 points. That means that, with 100 per cent, its quality of life must be perfect.

D

Traffic congestion · Air pollution · Crime · Street cleanliness · Race relations · Community spirit · Affordable housing · Cultural · Jobs/work · Health services · Young children · Teenagers · 20–40 year-olds · Elderly · Disabled · Shops · Schools · Public transport · Open space/parks · Sports and leisure amenities

2
a Complete table **E** to find the quality of life in your local area. Give each of the features a score of 1 to 5.

b Add up the scores to get a mark out of 100.

c Do you consider your local area to have a good, average or poor quality of life?

d In what ways does your local area have a good quality of life and in what ways does it have a poor quality of life?

e What would you do to improve the quality of life in your local area?

E

Facilities for … (Plenty = 5, Few/none = 1)		Access to … (Near = 5, Long way = 1)		Levels of … (Low = 5, High = 1)		Levels of … (High/good = 5, Low/poor = 1)	
Young children		Shops		Traffic congestion		Race relations	
Teenagers		Schools		Air pollution		Community spirit	
20–40 year-olds		Sports and leisure		Crime		Affordable housing	
Elderly		Open space/parks		Street cleanliness			
Disabled		Public transport					
		Health services					
		Jobs/work					
		Museums/library					

Summary

We can all help improve the environment and quality of life in the area in which we live.

How can the internet help?

Thanks to the **internet**, collecting information about a place has never been easier. There is really no limit to what you can find out about your local area, the United Kingdom and even the world. Indeed, the biggest problem is that so much information is available that it can take a long time to search through all the websites for the information you need.

The best way to overcome this problem is to know exactly what you want before you start. This way you can ask the right questions and waste less time searching. The Key Geography website should also make internet use much quicker and easier. It has a list of useful websites for every unit of the Key Geography course.

The two websites shown on these pages are a good place to start when collecting information about the local area. **Multimap** provides aerial photos and maps at a variety of scales (like those shown in **A**) for every location in the UK. The photos and maps can either be viewed alone or placed one on top of the other to help identify features.

For facts and figures try the government **National Statistics** website. It has information on employment, earnings, health, home ownership and many other topics. Typing in your postcode will give you information about your local area. You can easily extend your search to other areas and will also be given UK averages to compare with your own.

Other useful sources include local newspapers, many of which have websites of their own. These will have information about local issues such as education, job provision, the environment and planning proposals.

Aerial photo

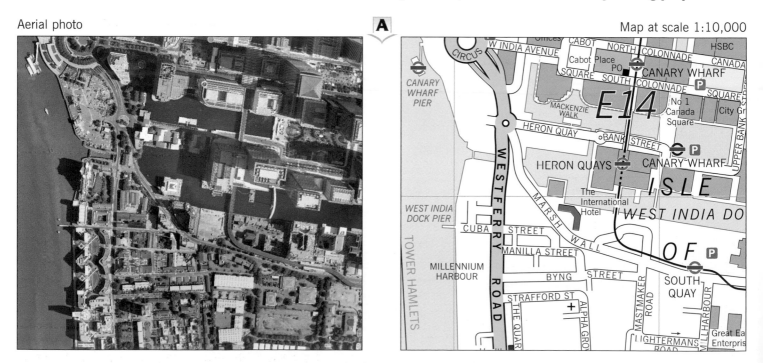

A

Map at scale 1:10,000

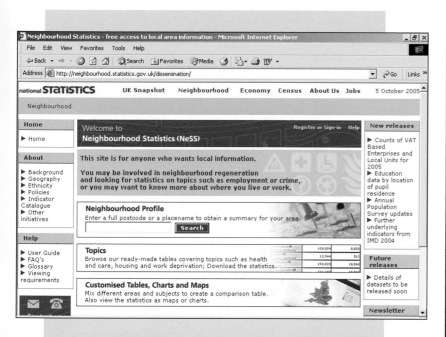

B

Activities

Go to: www.nelsonthornes.com

secondary/geography/key_geography.htm

1 a Log on to the Multimap website.

b Enter the postcode E14 which is London Docklands (see page 72).

c Change the scale to 1:10,000 and you should have a display like the one shown in **A**.

d Look at the maps and photos at several different scales to see what the area is like.

2 a Now log on to the National Statistics website (**B**, top).

b Click 'neighbourhood' at the top of the screen.

c Enter 'Tower Hamlets' for London Docklands. You should have a page like the one in **B** (bottom).

d Scroll down to 'Housing and Households' and you should have a table like the one in **C**.

3 a Describe what table **C** shows about house prices in Tower Hamlets compared with the rest of the country.

b Copy the table below and complete it using the census data in table **C**.

Types of housing		
	Tower Hamlets	**England & Wales**
Highest %	1	1
↑	2	2
↓	3	3
Lowest %	4	4

c Describe the main differences between housing in Tower Hamlets and the rest of the country.

4 Now repeat activities 1, 2 and 3 using maps and census data for the town or area where you live.

Summary

Some information about places may best be found on the internet. Using the Nelson Thornes website can make internet use much quicker and easier.

House types   **C**   **prices in £s**

	Tower Hamlets		England & Wales	
	Average price	Percentage of households living in this type of property	Average price	Percentage of households living in this type of property
Detached	220,688	1.0	178,806	22.8
Semi-detached	224,879	2.2	101,733	31.6
Terraced	228,726	13.0	89,499	26.0
Flat	204,568	83.6	120,184	19.2
All property types	208,367	100.0	119,436	100.0

This enquiry is concerned with your ability to use a range of resources, your understanding of what England and Wales is like, and what you have learned from this unit on the UK.

Your task is to plan a seven-day sightseeing tour of England (and Wales if you wish) for a pen-pal who lives in either the EU or the USA and is coming to the UK to visit you. Your tour should include visits to places that are both interesting and typical of your country. Your pen-pal is due to:

◆ arrive at either your local airport or one of London's airports on a Monday morning

◆ depart from London airport on the following Sunday afternoon.

Your task should be divided into three main parts.

1 You will need an introduction in which you explain what the enquiry is asking you to do.

2 You will have to collect various resources and explain how they will help you plan the tour.

3 You will need to describe the tour. This should be done in diary form. The diary should include reasons for your choice of places to be visited, and a reference as to how you will travel from one place to the next. This section should be illustrated with labelled maps and photos.

A

What are some of England's most interesting places?

1 Introduction

You should decide:

◆ which places your pen-pal is most likely to want to visit

◆ which places will give your pen-pal a good idea of what England is really like.

You should try to include visits to:

◆ a rural area that has attractive scenery – this could be mountains, lakes, a river valley or a stretch of coastline

◆ an urban area where there are lots of interesting places to visit

◆ an historic building, e.g. a castle or a cathedral or an historic site (for example a scene of a battle or famous event)

◆ a place with a lots of entertainment.

2 Planning the tour

This will allow you to use a wide range of resources. Useful resources might include:

◆ national bus and rail timetables

◆ rail, bus and car route maps

◆ CD-ROMs such as Encarta or AutoRoute Express

◆ the internet and the Nelson Thornes website

◆ holiday brochures.

3 Writing up the tour

Write up, in diary form, your proposed seven-day tour. You could set this out as in table **D** below.

◆ Draw a map to show your proposed route. Use different colours for different types of transport.

◆ Justify your choice of places that you plan to visit.

> ✔ **Remember**
> You must be practical in terms of time and money. You only have seven days and so cannot visit everywhere in the country. You also need to be careful with money – you cannot fly everywhere!

Follow-up work

◆ Compare your tour plans with those of others in your class. In what way are the tours:

  a similar

  b different?

◆ As a class you could then:

  a decide who produced the most varied and interesting route, and then

  b produce a combined effort, perhaps for a wall display.

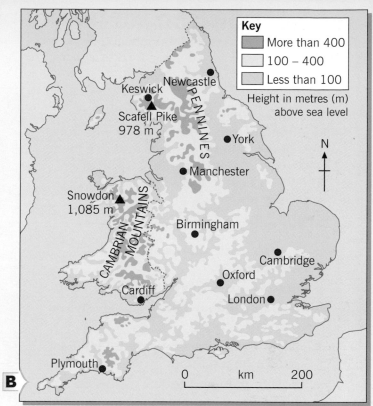

B

Key
- More than 400
- 100 – 400
- Less than 100

Height in metres (m) above sea level

Relief and places

C

BIRMINGHAM

181	CAMBRIDGE							
172	330	CARDIFF						
317	413	480	KESWICK					
192	96	248	501	LONDON				
142	258	309	192	128	MANCHESTER			
333	368	510	142	458	230	NEWCASTLE		
109	160	174	434	91	258	413	OXFORD	
325	469	261	632	386	461	662	312	PLYMOUTH
214	251	392	184	339	114	139	296	544	YORK

London to Oxford = 91 km

Differences between places on map **B**

D

	Places to be visited	Reasons for visit	Travel between places	Distance between places	Overnight stay
Day 1 Monday	Arrive at …				
Day 2 Tuesday					
Day 3 Wednesday					
Day 4 Thursday					
Day 5 Friday					
Day 6 Saturday					
Day 7 Sunday	Fly from London airport …				

How can we use maps?

What is this unit about?

The aim of this unit is to help you use, understand and enjoy maps. It will show you how to interpret maps and use them as a source of information. It will also show you how to locate places and find your way about.

In this unit you will learn:

◆ how to work out distance and direction

◆ how to use map symbols

◆ how to use four figure and six figure map references

◆ how height and shape of the land are shown on a map

◆ how to plan and follow routes on a map.

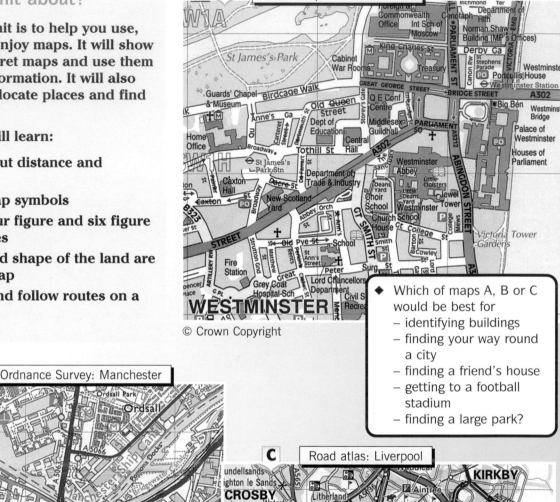

A Street map: Central London

© Crown Copyright

◆ Which of maps A, B or C would be best for
  – identifying buildings
  – finding your way round a city
  – finding a friend's house
  – getting to a football stadium
  – finding a large park?

B Ordnance Survey: Manchester

© Crown Copyright

C Road atlas: Liverpool

© Crown Copyright

Why is learning map skills useful to us?

Maps are useful to everybody.

◆ They give information, tell us where places are and help us find our way about.

◆ They show features and amenities of an area.

◆ They tell us how steep or flat the land is.

◆ They can help us find a friend's house or the best way to school, to shops and to a holiday destination.

Learning about maps can also help us to use, and get the best out of satellite and computer mapping systems such as:

◆ **Global Positioning Systems (GPS)**

◆ **satellite navigation (sat-nav)** and

◆ **Geographic Information Systems (GIS).**

An example of a computer mapping system that can be linked to a GPS unit is shown in **D**, **E** and **F** below.

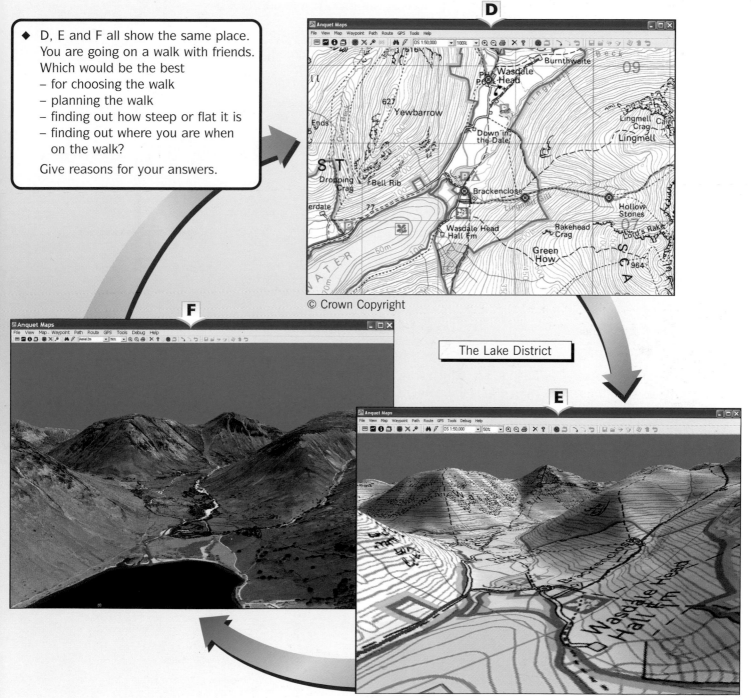

◆ D, E and F all show the same place. You are going on a walk with friends. Which would be the best
  – for choosing the walk
  – planning the walk
  – finding out how steep or flat it is
  – finding out where you are when on the walk?

Give reasons for your answers.

D

© Crown Copyright

The Lake District

E

F

© Crown Copyright

How can we show direction?

Maps show what things look like from above. They are very useful because they give information and show where places are. There are many different types of map. These include street maps, road maps, **atlas** maps and **Ordnance Survey** (OS) maps.

A **plan** is a type of map. Plans give detailed information about small areas. Places like schools, shopping centres, parks and leisure centres are shown on plans.

This section is about **direction**. The best way to show direction is to use the **points** of the compass. There are four main points. These are north, east, south and west. You can remember their order by saying 'Never Eat Shredded Wheat'.

Between these four main points there are four other points. These are north-east, south-east, south-west and north-west.

Most maps have a sign to show the **north** direction. If there is no sign the top edge of the map should be **north**.

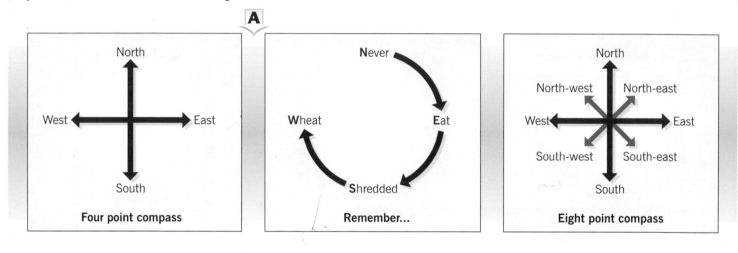

A

Four point compass

Remember...

Eight point compass

To give direction for a place you have to say which way you need to go to get there. The direction is the point of the compass *towards* which you have to go. Diagrams **B**, **C** and **D** show you how to give a direction.

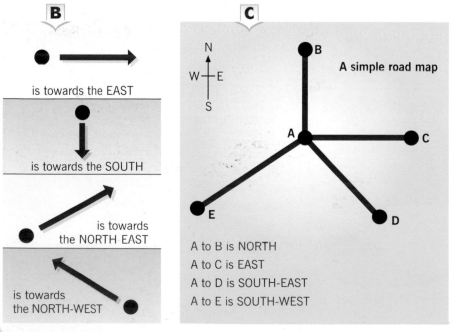

B

is towards the EAST

is towards the SOUTH

is towards the NORTH EAST

is towards the NORTH-WEST

C

A simple road map

A to B is NORTH
A to C is EAST
A to D is SOUTH-EAST
A to E is SOUTH-WEST

D

The kite is NORTH of the boat.
The ring is SOUTH of the ball.
The boat is WEST of the ring.
The flower is SOUTH-EAST of the boat.
The ring is NORTH-EAST of the flower.

Activities

1 Draw the compass in diagram **E** and label the unmarked points.

E

North-east

East

South-east

2 Copy these drawings and complete the sentences below them. The first one has been done for you.

B   D   E   G

A   C   F   H   I   J

B is north of **A**   **D** is . . . of **C**   **F** is . . . of **E**   **H** is . . . of **G**   **I** is . . . of **J**

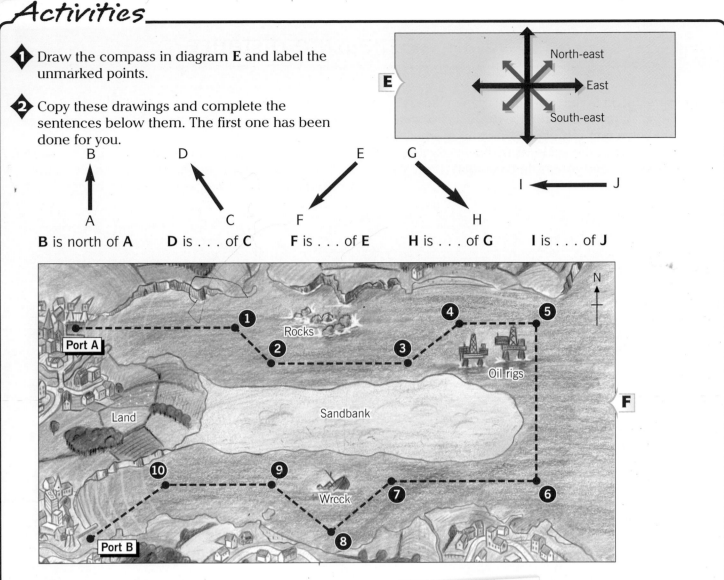

3 Study map **F** and give the following directions:

a from Port A to the rocks

b from the wreck to the oil rigs

c from the oil rigs to the rocks

d from the wreck to Port A

e from the rocks to the wreck.

4 a A ship has landed its cargo at Port A. It must go to Port B to reload. The course the ship must follow is shown by the dotted line on the map. Give the Captain compass directions to follow between each numbered point. Start like this: *Leave Port A. Go east to point 1. Go south-east ...*

b Imagine that the sandbank has been cleared to make ship movement easier. Work out the best course from Port B to Port A. Give compass directions to follow that course.

EXTRA

You will need to use the Ordnance Survey map of the Cambridge area for this question. It is on the inside back cover.

Look at the villages near the bottom of the map. Give the following directions:

a from Foxton to Whittlesford

b from Foxton to Newton

c from Great Shelford to Whittlesford

d from Great Shelford to Haslingfield

e from Haslingfield to Harston.

Summary

Maps are a good way of giving information and showing where places are. Direction can be described by using the points of the compass.

How can we measure distance?

A map can be used to find out how far one place is from another. Maps have to be drawn smaller than real life to fit on a piece of paper. How much smaller they are is shown by the **scale**. This shows you the **real** distance between places. In diagram **A** the scale line shows that 1 cm on the map is the same as 1 km on the ground. Every map should have a **scale line**.

Straight line distances are easy to work out. Diagram **A** shows how to measure the straight line, or shortest, distance between the church and the bridge.

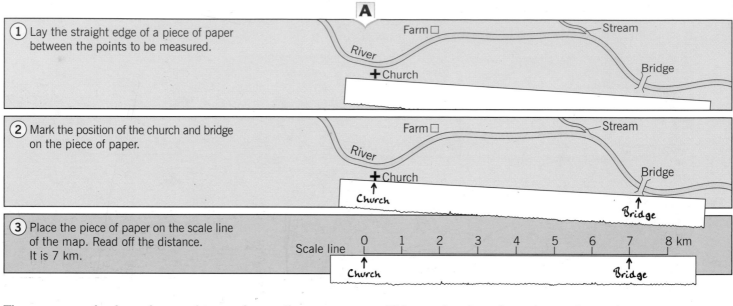

① Lay the straight edge of a piece of paper between the points to be measured.

② Mark the position of the church and bridge on the piece of paper.

③ Place the piece of paper on the scale line of the map. Read off the distance. It is 7 km.

The same method can be used to work out distances that are not straight lines. To measure these, divide the route into a number of sections and measure each one.

This can be done by using a piece of paper and turning it at each bend. Diagram **B** shows how to measure the distance from the church to the bridge, following the river.

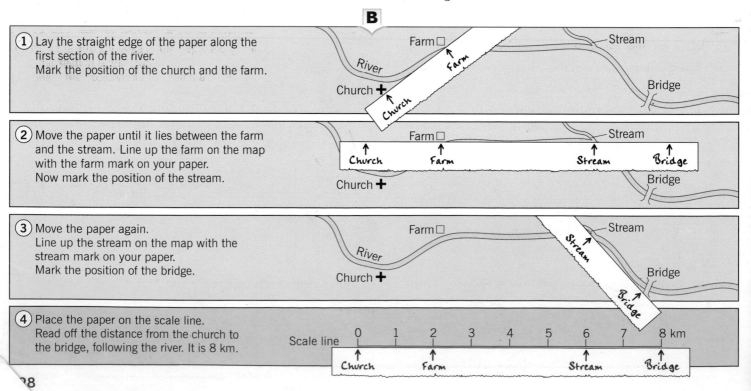

① Lay the straight edge of the paper along the first section of the river. Mark the position of the church and the farm.

② Move the paper until it lies between the farm and the stream. Line up the farm on the map with the farm mark on your paper. Now mark the position of the stream.

③ Move the paper again. Line up the stream on the map with the stream mark on your paper. Mark the position of the bridge.

④ Place the paper on the scale line. Read off the distance from the church to the bridge, following the river. It is 8 km.

Activities

1 Use the scale line from map **C** to give the lengths of these lines. Answer like this:

Line **a** is ... metres (m) in length.

a _____

b _____

c _____

d _____

2 Use map **C** and the scale line to give the straight line distance between the places below. Choose your answers from the following:

| 40 m | 80 m | 100 m | 120 m |

a Kate's house and the school

b Joanne's house and the post office

c Tim's house and the post office

d John's house and the garage

3 a Give the distance Joanne has to travel to school if she calls on Kate on the way.

b Give the distance John has to travel to school if he calls at the shop and post office first.

4 What is the distance around the duck pond if you walk on the footpath? Give your answer in metres (m).

5 You have been given a map and some instructions to help you find some hidden treasure. Follow the instructions to find out where it is.

Check the exact spot by sorting out the jumbled words in the treasure chest and choosing the correct answer.

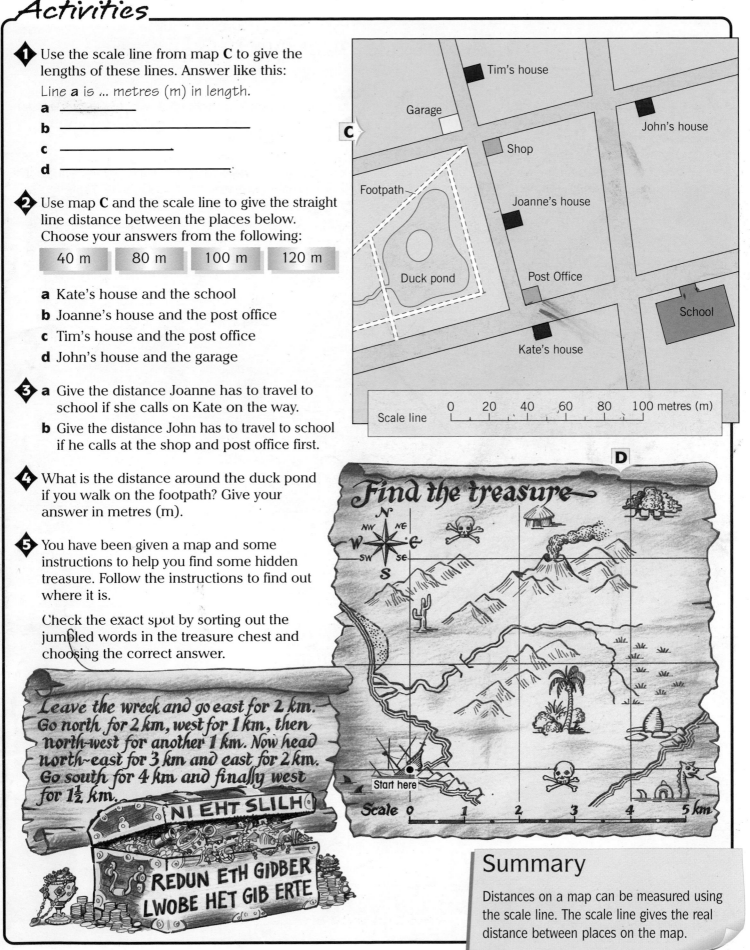

C

Tim's house

Garage

John's house

Shop

Footpath

Joanne's house

Duck pond

Post Office

School

Kate's house

Scale line
0 20 40 60 80 100 metres (m)

Find the treasure

N NE
NW
W · E
SW SE
S

Leave the wreck and go east for 2 km. Go north for 2 km, west for 1 km, then north-west for another 1 km. Now head north-east for 3 km and east for 2 km. Go south for 4 km and finally west for 1½ km.

NI EHT SLILH

REDUN ETH GIDBER LWOBE HET GIB ERTE

Start here

D

Scale 0   1   2   3   4   5 km

Summary

Distances on a map can be measured using the scale line. The scale line gives the real distance between places on the map.

How do we use map symbols?

A map must be clear and easy to read. There is always a lot to put on a map and it can easily become crowded. **Symbols** are used to save space and to make it easier to see things. Symbols may be small drawings, lines, letters, shortened words or coloured areas. The symbols used on a map are explained in a **key**.

If you are drawing your own map, you can make up your own symbols. They should be as simple as possible and look something like the feature they stand for. How would you show a postbox, a library or a football ground?

Sketch **A** and map **B** show the same street. The map has simplified the street scene. Only the main features of the street are shown and symbols are used to save space. The symbols are explained in the key.

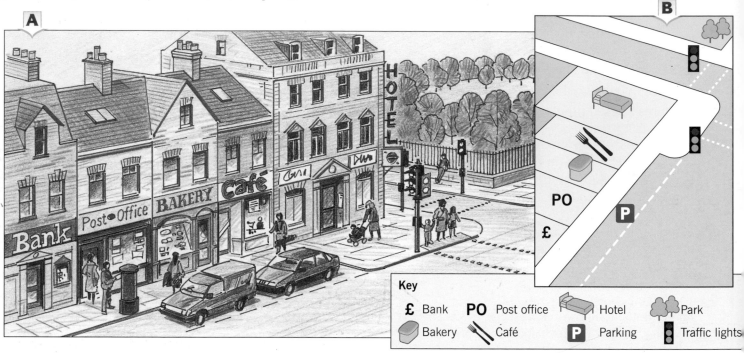

Key

£ Bank   PO Post office   Hotel   Park

Bakery   Café   P Parking   Traffic lights

The **Ordnance Survey** (OS) is responsible for mapping Britain. The OS produces very accurate maps that have a lot of information on them.

There is an Ordnance Survey map of the Cambridge area on the inside back cover of this book. The symbols used on that map are also shown on the inside back cover.

Look at the photos in **C**. They show some of the symbols used on Ordnance Survey maps. Which symbols could you work out without the answers being given?

Activities

1 Look at map **D**. It is part of the Ordnance Survey map of the Cambridge area on the inside back cover. It has been enlarged to make it easier to read. The scale has changed so the 4 cm on the map equals 1 km on the ground.

a Make a copy of table **E** below.

b Draw the symbols from map **D** in the correct columns of your table.

c Say what each symbol shows. You will need to use the key on the inside back cover. Some have been done for you.

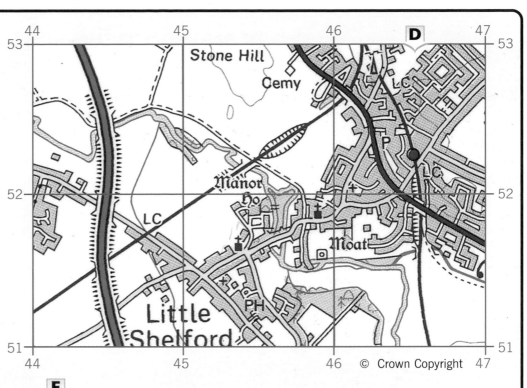

© Crown Copyright

E

Drawings	Lines	Abbreviations (letters/shortened words)	Coloured areas
\\\|// /||\\ = Embankment	‿‿‿ = Contour	Cemy = Cemetery	= Buildings

2 Make a larger copy of map **F**. It should be at least half a page in size. Using the Ordnance Survey symbols from the inside back cover, draw on the map the following information.

- There is a main road between Gorton and Bayhead, and a second class road between Bayhead and Asham. A minor road joins Asham and Gorton and goes on to Calder.

- A railway line runs from Asham to Gorton, to Calder and on to Bayhead. The station at Calder is closed but the others are open.

- Gorton has a church with a spire and a chapel. Bayhead has a church with a tower and a post office. Asham has a telephone box and a youth hostel.

- The spot height at Big Hill is 312 metres high. The land south of Big Hill is marshy.

- The River Bee rises to the north of Big Hill and flows into the sea at Calder. (Remember to use bridges.)

- There is a wood on the east coast.

F

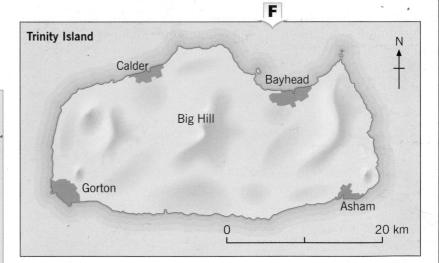

3 Draw a map of an island of your own. Use at least **15** different symbols. Name your towns, villages and other main features. Give your map a title.

Summary

Symbols are simple drawings that show things on maps. All maps have a key to explain the symbols.

What are grid references?

Maps can be quite complicated and it may be difficult to find things on them. To make places easier to find, a grid of squares may be drawn on the map. If the lines making up the grid are numbered, the exact position of a square can be given.

On Ordnance Survey maps these lines are shown in blue and each has its own special number. The blue lines form **grid squares**. **Grid references** are the numbers which give the position of a grid square. On these two pages you will learn about **four figure grid references**.

To *give* a grid reference is simple. Look at the grid in diagram **A** and follow these instructions to give the reference for the yellow square.

◆ Give the number of the line on the *left* of the yellow square – it is **04**.

◆ Give the number of the line at the *bottom* of the yellow square – it is **12**.

◆ Put the numbers together and you have a four figure grid reference. It is **0412**.

In the same way, the Picnic Square has a reference of 0313 and the Church Square is 0512.

What will be the grid reference for the Bridge Square the Tent Square?

To *find* a grid reference is also easy. Look at the grid in diagram **B** and follow these instructions to find grid square ⬚. 4237

◆ Go along the top of the grid until you come to **42**. That line will be on the *left* of your grid square.

◆ Go up the side of the grid until you come to **37**. That line will be at the *bottom* of your square.

◆ Now follow those two lines until they meet. Your square will be above and to the right of that point. There is a house in it.

What is in squares 4136 and 4037?

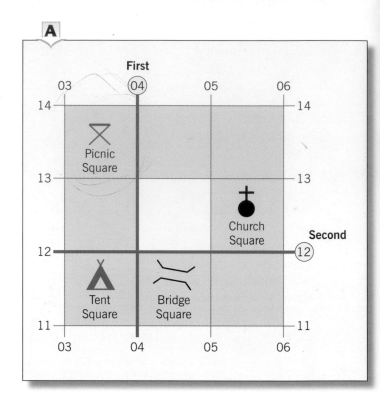

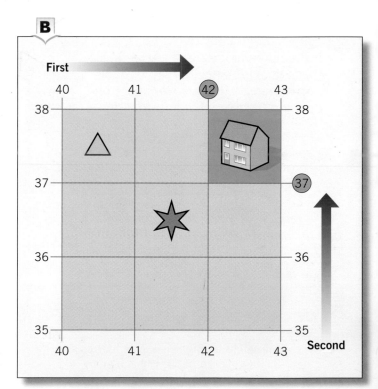

Activities

Look at map **C** of the British Isles. It shows some of the main towns, mountain areas and the three longest rivers. Use the map to answer the questions below.

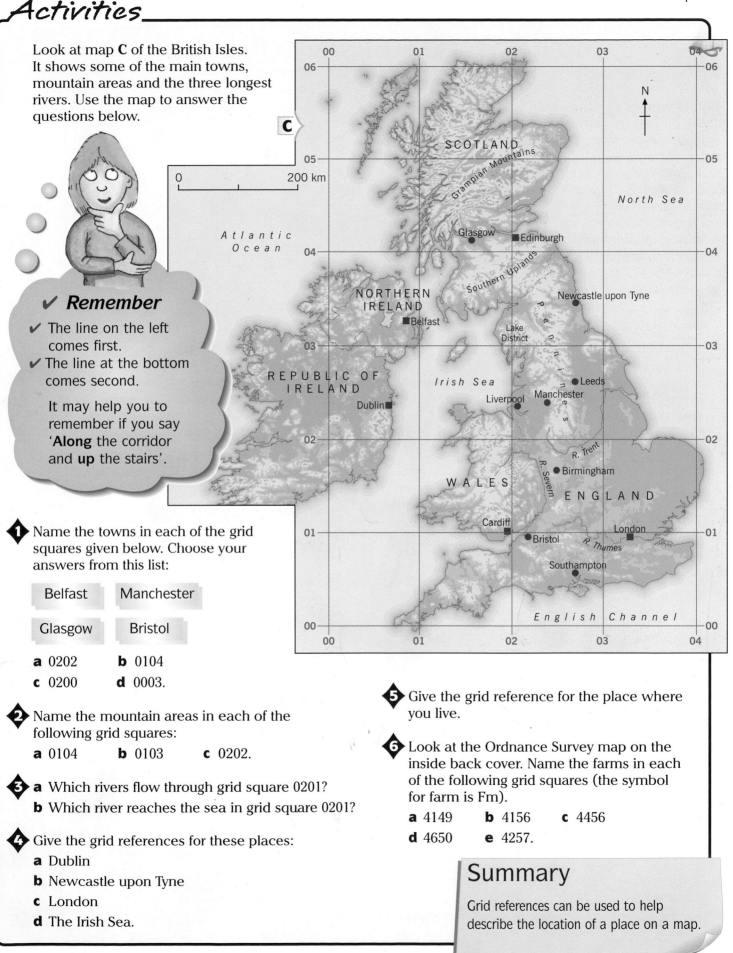

✔ **Remember**

✔ The line on the left comes first.
✔ The line at the bottom comes second.

It may help you to remember if you say '**Along** the corridor and **up** the stairs'.

1 Name the towns in each of the grid squares given below. Choose your answers from this list:

Belfast	Manchester
Glasgow	Bristol

a 0202   **b** 0104
c 0200   **d** 0003.

2 Name the mountain areas in each of the following grid squares:

a 0104   **b** 0103   **c** 0202.

3 a Which rivers flow through grid square 0201?
b Which river reaches the sea in grid square 0201?

4 Give the grid references for these places:

a Dublin
b Newcastle upon Tyne
c London
d The Irish Sea.

5 Give the grid reference for the place where you live.

6 Look at the Ordnance Survey map on the inside back cover. Name the farms in each of the following grid squares (the symbol for farm is Fm).

a 4149   **b** 4156   **c** 4456
d 4650   **e** 4257.

Summary

Grid references can be used to help describe the location of a place on a map.

133

How do we use six figure grid references?

Grid references are very useful in helping us to find places on maps. A four figure reference on an Ordnance Survey map equals an area on the ground of one square kilometre. This is quite a large area. To be more accurate we need to use a **six figure grid reference**. This pinpoints a place exactly to within 100 metres.

Look at the grid in diagram **A**. The six figure grid reference for the church is 045128. Follow these instructions and look at diagrams **B** and **C** to see how that reference is worked out.

◆ Give the number of the line on the *left* of the yellow square – it is **04**.

◆ In your head divide the square into tenths as shown in the grid in diagram **B**. Follow arrow **A** across the square. The church is about halfway across from the left. That puts it on the five-tenths line. Write down **5** after your number 04.

◆ You now have the first half of your six figure reference – **045**.

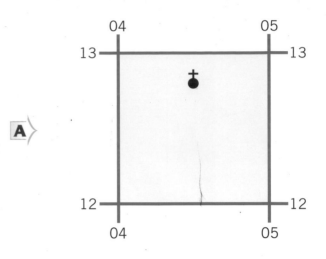

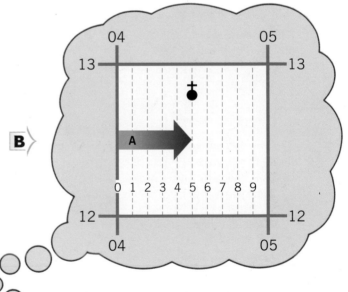

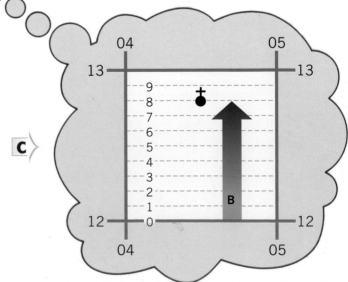

✔ **Remember**

✔ The numbers along the **bottom** come first.
✔ The numbers on the **left** come second.
✔ There must always be six figures.

◆ Now give the number of the line at the *bottom* of the yellow square – it is **12**.

◆ In your head divide the square into tenths as shown in the grid in diagram **C**. Follow arrow **B**. The church is over halfway up from the bottom. That puts it on the eight-tenths line. Write down **8** after your number 12.

◆ You now have the second half of your six figure reference – it is **128**.

◆ Put the two halves together and you have **045128**.

Activities

Look at map **D**. The 'tenths' lines have been added to help you with activities **1**, **2** and **3**. Check your references ...

- The village of Eldon is in grid square 1623.
- The Mill is at reference 166256.
- Dingle Farm is at reference 170238.

1 Copy and complete the sentences below. Use the correct answer from the brackets.

  a At 168245 there is a (church, post office, farm).

  b At 165257 there is a (telephone, school, bridge).

  c At 175233 there is a (farm, lake, level crossing).

  d At 177244 there is a (station, wood, roundabout).

2 Give the six figure grid reference for each of the following:

  a Eldon post office

  b Causey railway station

  c Padley school

  d Burr Wood picnic site.

3 a Follow these directions for a pleasant walk:

> Start at 170238. Walk down to 173237. Turn left and go to 177244. Go along the road to 171248. Follow the path to 178257. Turn left and finish your walk when the path reaches the road.

  b Name the place where you finished your walk. Give its six figure grid reference.

  c Where would you have stopped for lunch?

  d How many churches did you pass on the way? Give their six figure grid references.

4 You will need to use the Ordnance Survey map of the Cambridge area for this question. It is on the inside back cover.

  a Make a copy of table **E**.

  b Use the map to complete table **E**. The missing symbols, meanings and references are given in diagram **F**.

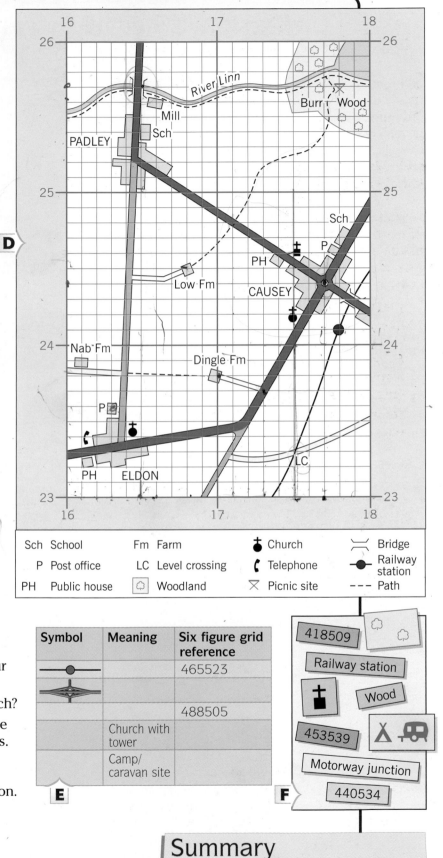

D

Sch School   Fm Farm   ✝ Church   ⎵ Bridge
 P Post office   LC Level crossing   ☎ Telephone   ●— Railway station
PH Public house   🌳 Woodland   ✕ Picnic site   - - - Path

Symbol	Meaning	Six figure grid reference
—●—		465523
◆		
		488505
	Church with tower	
	Camp/ caravan site	

E

F

418509

Railway station

✝

Wood

453539

🏕 🚐

Motorway junction

440534

Summary

Six figure grid references can be used to give the exact position of a place on a map.

How is height shown on a map?

The land around us is seldom flat. There are nearly always differences in height and differences in slope. Sometimes slopes may be gentle and at other times steep. There may be hills, mountains and valleys or areas that are quite level. The word **relief** is used by geographers to describe the shape of the land.

Map makers have to find ways of showing relief and height. How they do this is shown on the next four pages.

Look at sketch **A**. How can height on the island be shown on a flat piece of paper? Height is usually measured from sea level in metres. This can then be shown on a map in three different ways. These are by using **spot heights, layer colouring** and **contours**.

A

Spot heights   **B**

These give the exact height of a point on the map. They are shown as a black dot and each one has a number next to it. The number gives the height in metres. A **triangulation pillar** is also used to show height. These are drawn as a dot inside a blue triangle on the map.

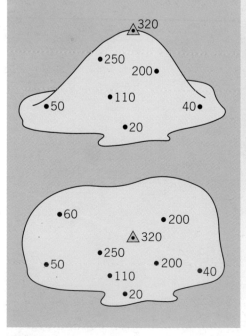

Layer colouring   **C**

This can also be called **layer shading**. Areas of different heights are shown by bands of different colours. Brown is usually used for high ground, and green for low ground. There always needs to be a key. Layer colouring is used in atlases to show height.

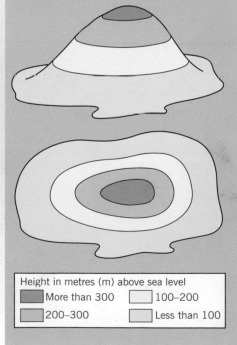

Height in metres (m) above sea level	
▉ More than 300	☐ 100–200
▨ 200–300	☐ Less than 100

Contours   **D**

Contours are lines drawn on a map. They join places which have the same height. They are usually coloured brown. Most contours have their height marked on them but you may have to trace your finger along the line to find it. Sometimes you will have to go to the contour above or below to get the height. Heights are given in metres.

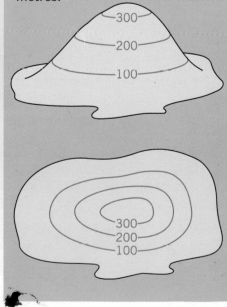

Activities

1 a Copy out and complete crossword **E** using the clues below.

b When you have finished, give the meaning of the downword in the orange squares.

Clues

1 Lines that join places of the same height.
2 Height at one place.
3 This can be gentle or steep.
4 Measured from sea level.
5 Colouring to show height.
6 A level area with no slope.

E ↓   Downword ↓

1► 2► 3► 4► 5► 6►

2 Look at map **F** of England and Wales. The map uses layer colouring to show height. The letters mark land at different heights.

a Which letters mark lowland areas under 100 metres?

b Which letters mark land between 100 and 500 metres?

c Which letters mark land above 500 metres?

3 Use map **F** to answer these questions.

a The highest mountain in England is Scafell Pike and the highest mountain in Wales is Snowdon. What colour are they shaded?

b The Pennines are an area of high land in the centre of northern England. How high are they?

c The Cotswolds and Chilterns are hills in the south of England. What height are they?

d What height is the area where you live?

F

Key

Height in metres above sea level

- Over 500
- 200–500
- 100–200
- Below 100

N

Scafell Pike 978 m

Snowdon 1,085 m

PENNINES

ENGLAND

WALES

COTSWOLDS

CHILTERNS

0   100 km

EXTRA

Look at the Ordnance Survey map on the inside back cover.

1 Give the heights above sea level of the following:

a the contours in grid squares 4852 and 4450

b the spot heights in grid squares 4151 and 4754

c the triangulation pillar in grid square 4051.

2 Look at Rowley's Hill in grid square 4249. Draw the pattern of contours and the triangulation pillar. Write in any heights that are given.

Summary

There are three main methods of showing height on maps. These are spot heights, layer colouring and contours.

How do contours show height and relief?

Lines on a map that join places of the same height are called **contours**. Contours show the height of the land and what shape it is. The shape of the land is called **relief**. The difference in height between contours is chosen by the map maker. On most Ordnance Survey maps they are drawn at every 10 metres. This difference in height is called the **contour interval**. Several contours together make up a pattern. By looking carefully at these patterns you can work out how steep the slopes are and what shape the land is.

Contour lines are drawn on maps by map makers. You cannot see them on the ground. In diagram **A** the contours have been drawn on the main sketch. You will see that they make up different patterns. An important thing to remember is that:

◆ *the closer the contour lines are together, the steeper the slope will be.*

A

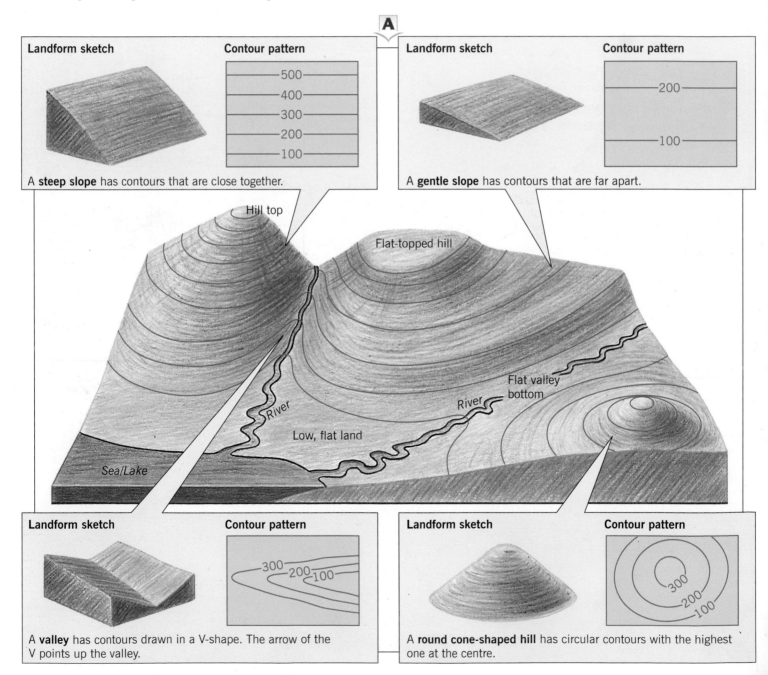

Landform sketch   **Contour pattern**

500
400
300
200
100

A **steep slope** has contours that are close together.

Landform sketch   **Contour pattern**

200

100

A **gentle slope** has contours that are far apart.

Hill top

Flat-topped hill

River

Flat valley bottom

River

Low, flat land

Sea/Lake

Landform sketch   **Contour pattern**

300
200
100

A **valley** has contours drawn in a V-shape. The arrow of the V points up the valley.

Landform sketch   **Contour pattern**

300
200
100

A **round cone-shaped hill** has circular contours with the highest one at the centre.

Activities

1 From map **B** give the heights of the following places. Choose your answers from those in the brackets.

  a The highest point is (22, 48, 52, 40, 60) metres.

  b Place **E** is (8, 42, 30, 20, 16) metres.

  c Place **B** is (30, 20, 26, 46, 34) metres.

  d Place **A** is (15, 10, 34, 6, 21) metres.

  e Place **D** is (28, 10, 12, 22, 8) metres.

2 Look at map **B** and say if the following statements are TRUE or FALSE.

  a **E** and **F** are at the same height.

  b **D** is higher than **F**.

  c **B** is higher than **E** but lower than **C**.

  d **A** is the lowest place marked with a letter.

  e **D** to **C** is steeper than **A** to **B**.

3 The photos in **C** show some landscape features.

  a Draw a simple contour pattern for each of the photos.

  b Write a description of the feature next to each of your drawings.

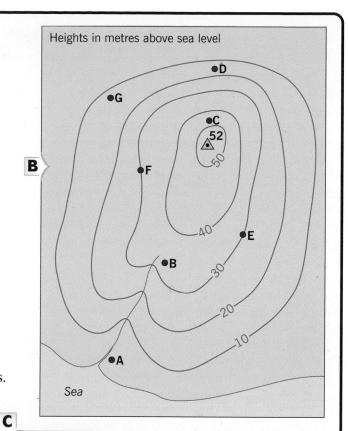

Heights in metres above sea level

B

Sea

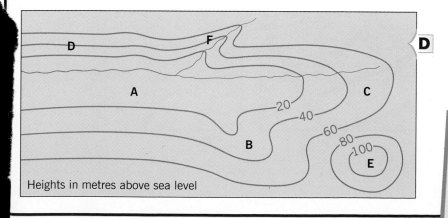

C

4 Look at the six letters on map **D**. Match the letters to each of the following:

  1 A gentle slope   **4** A flat valley floor

  2 A steep slope   **5** A valley with a stream

  3 A hill top   **6** A valley without a stream.

D

Heights in metres above sea level

Summary

Contour lines are a good way of showing height and relief on a map. Contours that are close together show steep slopes. Contours that are far apart show gentle slopes.

How can we describe routes?

Maps show what things look like from above. They have a lot of information on them. You can use this to describe where places are and work out what may be seen there. Maps are also useful for describing routes between places. These two pages show how to describe routes and places from Ordnance Survey maps.

Paul lives in Foxton. He writes letters to a friend called Chris. Part of one of his letters is shown in **A**. It describes where he lives and a walk he often takes. Map **B** shows the area, which is near Cambridge. See if you can recognise the things Paul talks about. Can you follow the route?

Paul's description was good. He first described the area in general and then mentioned both the **physical** and **human** features. These are labelled on map **B**. When describing a place or a route, there is no need to try to include everything, but you must be very accurate.

A

Dear Chris,
I live in the small village of Foxton. It has a church, Post Office and Public House. The area around here is open countryside and mainly flat. I often take my dog for a walk across to Newton which is 2½ km away. We follow a path across fields until we reach a stream and some trees. Rowley's Hill is to the north. It is 50m high and gently sloping. We then follow the stream and trees until the path becomes a narrow road. The road passes a church and some gardens belonging to the Manor before it reaches Newton. My walk usually ends at the Post Office.

Rose Cottage
Foxton

B

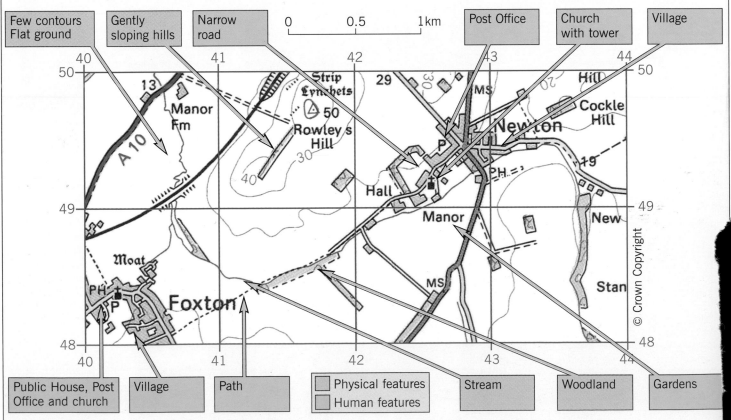

Few contours Flat ground	Gently sloping hills	Narrow road		Post Office	Church with tower	Village

Public House, Post Office and church | Village | Path | □ Physical features □ Human features | Stream | Woodland | Gardens

Activities

1 Look carefully at photo **C**.

   a List **three** physical features and **three** human features on it.

   b Write a brief description of the town.

2 Look at Newton on map **B**.

   a List the physical features and human features in and around the village.

   b Imagine that you own a cottage in the village that you want to rent out. Write a brief description of the village and surrounding area to advertise the cottage.

3 Map **D** shows part of the area covered by the OS map on the inside back cover.

   a Imagine that you have arrived by train at Great Shelford station and have to go to Hauxton post office. Some of the features that you will pass on your route are shown in diagram **E**. Write out the features in the order you would pass them. Begin with the station.

   b What is the distance from the station to the post office? Give your answer in kilometres.

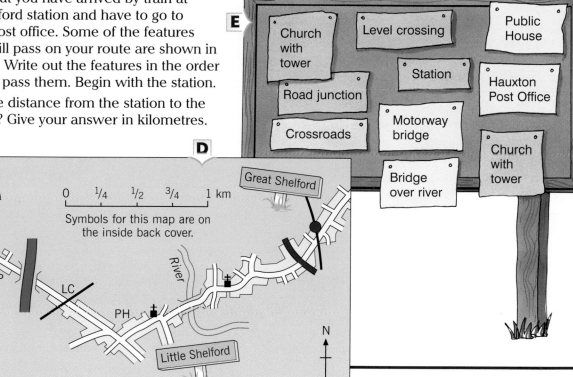

EXTRA

Use the Ordnance Survey map on the inside back cover for these questions.

1 Follow this route:
Start at Barton (4055).
Go on the A603 to junction 12.
Travel by motorway to junction 11.
Follow the A10 (T) south-west for 2 km.
Turn south-east down the B1368.
Stop after 2.5 km. Where are you now?

2 a Describe the route you would follow by road from Haslingfield (4052) to Grantchester.

   b Describe the village of Grantchester and the surrounding area. Mention both physical and human features.

Summary

Maps can be used to describe routes and places. Accuracy is very important when describing things.

Glossary and Index

A

Accessibility How easy a place is to get to. *76*

Anticyclone A weather system with high pressure at its centre. *32*

Aspect The direction which a slope or house faces. *26*

Atmosphere The air around the earth. *6–7*

B

Beaufort scale A scale for measuring wind speed using things like smoke and trees. *24*

By-pass A road built around a busy area to avoid traffic jams. *84–85*

C

Central business district (CBD) The middle of a town or city where most shops and offices are found. *70, 76*

Climate The average weather conditions of a place. *7, 28, 36–39, 107*

Clouds Masses of condensed water droplets suspended in the atmosphere. *25, 30*

Communications The ways in which people, goods and ideas move from one place to another. *8–9, 80–83*

Condensation The process by which water vapour changes to liquid water when cooled. *30, 42–43*

Congestion Overcrowding on roads causing traffic jams. *76, 80, 83–84*

Contour A line drawn on a map to join places at the same height above sea level. *136, 138, 140*

Contour interval The difference in height between contours on a map. *138*

Convectional rain Rain that is produced when air rises after being warmed by the ground. *30–31*

Corner shop Small shops in inner city areas selling things which people need every day. *74*

D

Deforestation The cutting down or burning of trees to clear large areas of land. *47*

Depression A weather system with low pressure at its centre. *34–35*

Direction Shown on a map by the points of the compass. *25, 126*

Dispersed settlement Several farms or buildings spread out over a wide area. *64–65*

Drought A long spell of dry weather. *7, 107*

E

Earthquake A movement or tremor of the earth's crust. *88–93, 98–99*

Economic geography Is about industry, jobs, earning a living and wealth. *8–9, 112*

Ecosystem A community of plants and animals that live together in the environment. *7*

Embankment A raised river bank to prevent flooding. *50, 54–56*

Emergency relief/aid Immediate help that is needed following a disaster. *97*

Emigrant A person who leaves a country to live and work in another country. *110*

Environment The natural or physical surroundings where people, plants and animals live. *5, 10, 21, 86–87*

Ethnic groups People with a similar culture, background and way of life. *111*

European Union A group of 25 European countries working together for everyone's benefit. *104*

Evaporation The process by which liquid water changes to water vapour when it is warmed. *42–43*

F

Fieldsketch A labelled sketch drawn outside of the classroom. *13*

Flash flood A sudden and unexpected flood that can cause much damage. *49*

Flood The flow of water over an area that is usually dry. It may be a river flowing over flat land beside it, or the sea covering a low-lying coastal area. *46–59*

Flood prevention scheme A plan to try to stop flooding by either rivers or the sea. *51, 54–57*

Four figure grid reference A group of four figures to help find a square on an Ordnance Survey map. *132*

Frontal rain When warm air has to rise over cold air in a depression. *31, 34*

Function The main purpose of a town or parts of a town. Functions include residential, industrial, commercial and recreational. *70*

G

Graphs Diagrams showing information in a pictorial way, e.g. how two variables such as population growth and time may be related. *16–17*

Grid square A square on a map representing an area on the ground. *132*

Groundwater Fresh water stored in rocks and the soil. It may pass slowly through the rocks and soil back to the sea. *42–43*

Hazard A natural danger to people and their property. Hazards include earthquakes, gales, drought and floods. *7, 46–59, 88–101, 107*

Height How high or low a place is, measured in metres or feet above sea level. *136, 138*

High and low order These are goods sold in a shop. High order goods cost a lot but are not bought very often, e.g. furniture. Low order goods cost less but are bought more often, e.g. food. *74*

Human geography Where and how people live. *4–5, 8–9, 20, 74, 108*

Immigrant A person who arrives in a country with the intention of living there. *110–111*

Inner city An area of factories and old houses next to the city centre. *70–73*

Internet shopping Ordering goods on the internet from the comfort of your own home and having them delivered to your door. *78–79*

Isobar A line on a map joining places with the same atmospheric pressure. *34*

Key A list of signs and symbols on a map or diagram with an explanation of what they mean. *130*

L

Land use Describes how the land in towns or the countryside is used. It includes housing, industry and farming. *70, 72, 116*

Landforms Natural features formed by rivers, the sea, ice and volcanoes. *6–7*

Landscape The scenery or appearance of an area. It includes both physical and human features. *12, 140*

Latitude This says how far north or south a place is from the equator. *14–15*

Layer colouring A method of showing height on a map by using colours. *136*

Linear settlement Buildings spread out in a line beside a main road, a railway or a river. *64–65*

Longitude This says how far east or west a place is from the Greenwich Meridian. *14–15*

Long-term Problems or assistance that last several months or years. *94, 97*

Map A drawing which shows part of the earth's surface from directly above, at a reduced scale. *14–15, 126–141*

Meteorology The study of the weather. *24–25*

Microclimate The climate of a small area. *26–27*

Migration The movement of people from one place to another to live or work. *8, 110–111*

Multicultural society A society where people with different beliefs and traditions live and work together. *110*

National Parks Areas of scenic beauty which are protected so that people can enjoy open air recreation. *11*

North Atlantic Drift A warm ocean current that brings mild conditions to the west of Britain in winter. *28*

Nucleated settlement Buildings that are grouped closely together. *64–65*

Ordnance Survey The official government organisation responsible for producing maps in the UK. *65, 116, 131, 140, inside back cover*

Pattern How things like settlements and shops are spread out over an area of land. *64–65, 70, 138*

Physical geography Natural features and events on earth. It includes landforms and weather. *4–7, 20, 106–107*

Place An area of the earth's surface. It can vary in size from a desk in a classroom to a city or a continent. *12–15*

Plan A detailed map of a small area. *126*

Plates Large sections of the earth's crust. *91*

Points of the compass A method of giving direction using north, south, east, west, etc. *126*

Pollution Noise, dirt and other harmful substances produced by people and machines which spoil an area. *10–11, 80, 83*

Population The people who live in an area. *8, 16, 110–111*

Precipitation Water in any form which falls to earth. It includes rain, snow, sleet and hail. *24, 30–31, 37, 42–43, 107*

Pressure Atmospheric pressure is the weight of air pressing down on the earth's surface. *32, 34*

Primary activities Jobs that involve the collecting and using of natural resources, e.g. farming, fishing, mining and forestry. *9, 112*

Public transport Transport provided to the public and available to everyone, e.g. buses, trains, etc. *82–83*

Quality of life How content people are with their lives and the environment in which they live. *9, 86–87, 112, 118–119*

Refugees People who have been forced to move from an area where they live, and have been made homeless. *111*

Relief The shape of the land surface and its height above sea level. *136, 138*

Relief rain Rain caused by air being forced to rise over hills and mountains. *30*

Reservoir An artificial lake used to store water. *49, 56*

Resources Things which are useful to people. They may be natural like coal and iron ore, and of value like money and skilled workers. *10*

Ribbon developments Settlements that have a long narrow shape beside a road or a river. *64–65*

River basin An area of land drained by a river and its tributaries. *6, 44*

River channel Where a river flows. It has a bed and two banks. *44*

River mouth The end of a river where it enters the sea or a lake. *44*

River source Where a river begins. *44*

Satellite image A photo taken from a satellite orbiting in space and sent back to earth. *114*

Scale The link between the distance on a map and its real distance on the ground. *128*

Secondary activities Where natural resources are turned or manufactured into goods that we can use, e.g. cars, computers. *9, 112*

Service activities Jobs that provide a service for people, e.g. teaching, nursing and shopping. *9, 112*

Settlement A place where people live. *8, 62–73*

Shopping malls Shopping areas which are under cover and protected from the weather. *76*

Short-term Problems or assistance that last just weeks or a few months. *94, 97*

Site The actual place where a settlement first grew up. *62–63*

Six figure grid reference A group of six figures used to give an exact position on a map. *134*

Spot height A point on a map with a number giving its height above sea level in metres. *136*

Standard of living How well off a person or a country is. *112*

Stores Part of the water cycle where water is held in reserve in the sea, on land or in the air. *42, 44, 56*

Suburbanised village A village with many new buildings added to it. *66*

Suburbs A zone of housing around the edge of a city. *70–71*

Surface water Water which lies on top of, or flows over, the ground. *42–43, 46*

Symbol A simple drawing or sign used to give information and save space on a map. *24–35, 130–131, inside back cover*

Temperature A measure of how warm or cold it is. *24, 26–28, 37, 107*

Tertiary activities Jobs that provide a service for people, e.g. teaching, nursing and shopping. *9, 112*

Transfers Part of the water cycle when water in various forms moves between the sea, the land and the air. *42, 44*

Transpiration The process by which water from plants changes into water vapour. *42–43*

Triangulation pillar A concrete pillar used by surveyors to find the exact height and position of a place. *136*

Tributary A small river which flows into a bigger river. *44*

Tsunami A large wave caused by an undersea earthquake. *88–101*

Urban An area of land which is mainly covered in buildings. *9, 68–70, 80–83*

Urban model The pattern of land use in a town. *70*

Urbanisation The growing proportion of people living in urban areas. *47*

Visibility The distance that can be seen. *25*

W

Water cycle The never-ending movement of water between the sea, the land and the air. *42–43*

Watershed The boundary between two river basins. *44*

Weather The day-to-day condition of the atmosphere. It includes temperature, rainfall and wind. *7, 24–39*